In the Dark

SOUTH ASIA DEVELOPMENT FORUM

In the Dark

How Much Do Power Sector Distortions Cost South Asia?

FAN ZHANG

South Asia Development Forum

Home to a fifth of mankind, and to almost half of the people living in poverty, South Asia is also a region of marked contrasts: from conflict-affected areas to vibrant democracies, from demographic bulges to aging societies, from energy crises to global companies. This series explores the challenges faced by a region whose fate is critical to the success of global development in the early 21st century, and that can also make a difference for global peace. The volumes in it organize in an accessible way findings from recent research and lessons of experience, across a range of development topics. The series is intended to present new ideas and to stimulate debate among practitioners, researchers, and all those interested in public policies. In doing so, it exposes the options faced by decision makers in the region and highlights the enormous potential of this fast-changing part of the world.

Contents

Boxes

Figures

Maps

Tables

Foreword

The countries of South Asia still need to expand electricity access and ensure electricity reliability. Looking forward, they will also have to satisfy the electricity demand of their fast-growing economies. More than 250 million people in the region still live without access to electricity—roughly a quarter of the global unserved population. Several of the countries in South Asia also face electricity shortages, leading to frequent power shedding. On average, per capita electricity consumption in the region is less than a quarter of the world average. As South Asia continues its growth trajectory and more people are connected to the grid, demand for electricity is set to increase rapidly over the coming decades. India alone is expected to account for 30 percent of growth in global energy demand between now and 2040.

Expanding and improving electricity services is imperative for economic growth and poverty alleviation in South Asia. The World Bank is helping countries in South Asia to meet their energy needs through direct investments, technical assistance and budget support. Total lending commitment to the region for energy projects reached US$8.6 billion at the end of fiscal 2018. The focus has been on providing low-carbon options for energy access, such as increasing the use of renewable energy and encouraging more efficient use of energy. The World Bank also supports individual countries' reform agendas, particularly those focused on enabling the creation of markets and improving sector governance. In addition, it encourages and facilitates regional efforts to promote greater cross-border trade of electricity.

Although large investments are urgently needed to plug energy gaps, reforms that address policy distortions in the energy sector could play a big part in making the best use of existing facilities, avoiding waste, attracting private investment, and promoting the shift toward a cleaner energy mix.

In support of a greater prioritization of reforms, the report presents an integrative analysis of energy sector distortions at different stages of electricity supply in the three largest countries in South Asia: Bangladesh, India, and Pakistan. Using a rigorous analytical framework and new microeconomic data, the analysis estimates how various types of distortions affect economies and social outcomes. The range of distortions considered is broad, encompassing the misallocation of fuel supply, inefficiencies in generation, high losses in distribution, and inadequate pricing of emissions from fossil fuel–based electricity generation.

New insights are gained by relying on two important methodological innovations. First, the analysis goes beyond looking at just fiscal costs, evaluating the impact of distortions from a welfare perspective. Rather than the cost of subsidies, the report assesses the loss of consumer welfare and producer surpluses, as well as the environmental and social costs. Second, the report adopts a broad definition of the power sector. Instead of focusing exclusively on generation, transmission and distribution, the analysis covers the entire supply chain of power supply, from upstream fuel supply to downstream access and reliability.

The report finds that the full cost of distortions in the power sector is far greater than previously estimated based on fiscal costs alone. The estimated total economic cost is 4–7 percent of GDP in Bangladesh, India, and Pakistan. Some of the largest costs are upstream and downstream.

The report also shows that countries in South Asia can reap huge economic gains from energy sector reforms, and along the way offers important insights on the implementation of these reforms. For example, a narrow focus on liberalizing the price of electricity should be avoided because, in the absence of other reforms, the market equilibrium is highly inefficient. It also appears that, without fundamental changes in incentives, corporatizing power utilities does not guarantee substantial improvements in their operation. And ensuring universal access to electricity without ensuring a reliable power supply amounts to a missed opportunity, because the benefits from electrification crucially depend on households and firms getting a sufficient level of services.

Through policy reforms, institutional development and infrastructure investments, South Asia can address energy supply challenges and cement a path to sustainable development. The World Bank stands ready to support the countries in South Asia in these efforts.

Hartwig Schafer
Vice President
South Asia Region
The World Bank

Acknowledgments

This book was prepared by a team led by Fan Zhang (Senior Economist) under the guidance of Martin Rama (Chief Economist) and Annette Dixon (Vice President) of the South Asia Region of the World Bank. The book was a collaborative effort of the Office of the Chief Economist and the Energy and Extractives Unit of the South Asia Region. Core team members included Umul Awan, Amol Gupta, Md. Iqbal, Fatima Najeeb, Mohammad Saqib, and Weijia Yao. Miklos Bankuti, Jagabanta Ningthoujam, and Jiarui Wang provided research assistance. Ron Chan, Corbett Grainger, Edward Manderson, Paolo Mastropietro, Brian Min, Ashish Rajbhandari, Pablo Rodilla, Hussain Samad, Filipe Lage de Sousa, Kai Sun, and Galina Williams contributed to background papers of the book. Barbara Karni, Alison Strong, and Sabra Ledent edited the book. Neelam Chowdhry provided timely administrative support. Aziz Gokdemir and Jewel McFadden oversaw the publication process.

Martin Rama provided invaluable insights during the preparation process. Julia Bucknall and Demetrios Papathanasiou provided substantial support and encouragement to the team.

Vivien Foster, Sheoli Pargal, Richard Spencer, and Michael Toman were the peer reviewers. They provided rich feedback throughout the preparation process.

The team is also grateful to other colleagues and experts for their helpful comments and inputs, in particular Anjum Ahmad, Partha Bhattacharyya, Yann Doignon, Marianne Fay, Alexander Anthony Ferguson, Virgilio Galdo, Elena Karaban, Jie Li, Yue Li, Gladys Lopez-Acevedo, Muthukumara Mani, Joe Qian, Abdul Wajid Rana, Nandita Roy, Eri Saikawa, Simon J. Stolp, Jari Vayrynen, and Salman Zaheer. The team apologizes to anyone inadvertently overlooked in these acknowledgments.

The team thanks the Partnership for South Asia Trust Fund, the Energy Sector Management Assistance Program, the South Asia Umbrella Facility for Gender Equality Multi-Donor Trust Fund, and the World Bank's Multi-Donor Trust Fund for Trade and Development for their generous financial support.

About the Author

Fan Zhang is a Senior Economist in the Office of the Chief Economist of the South Asia Region at the World Bank. Previously she worked in the Europe and Central Asia Region as a Senior Energy Economist for the Energy and Extractives Global Practice of the World Bank. She has led both lending and advisory programs and published in the areas of energy and environmental economics, economic growth, and climate change. Before joining the World Bank Group, she was an Assistant Professor of Energy Economics and Policy at Pennsylvania State University. She has a PhD from Harvard University.

Abbreviations

ARDL	autoregressive distributed lag model
BPDB	Bangladesh Power Development Board
CIL	Coal India Limited
DMSP-OLS	Defense Meteorological Satellite Program's Operational Linescan System
DSM	Deviation Settlement Mechanism
GCV	gross calorific value
GDP	gross domestic product
Gg	gigagram
GJ	gigajoule
GW	gigawatt
IFC	International Finance Corporation
IPP	independent power producer
kcal	kilocalorie
kV	kilovolt
kWh	kilowatt-hour
LNG	liquefied natural gas
LPG	liquefied petroleum gas
MMBtu	million British thermal units
MMcfd	million cubic feet per day
MSE	micro- and small enterprise
MW	megawatt
NEPRA	National Electric Power Regulatory Authority (Pakistan)
NOAA	National Oceanic and Atmospheric Administration (United States)
OECD	Organisation for Economic Co-operation and Development
OGRA	Oil and Gas Regulatory Authority (Pakistan)
$PM_{2.5}$	fine particulate matter 2.5 micrometers in diameter
PRs	rupees (Pakistan)

PSI	power supply irregularity
REB	Rural Electrification Board (Bangladesh)
RGGVY	Rajiv Gandhi Grameen Vidyutikaran Yojana (India)
Rs	rupees (India)
SAIFI	System Average Interruption Frequency Index
SAIDI	System Average Interruption Duration Index
SCCL	Singareni Collieries Company Limited (India)
SNGPL	Sui Northern Gas Pipelines Limited (Pakistan)
SSGC	Sui Southern Gas Company Limited (Pakistan)
TE	technical efficiency
TFP	total factor productivity
Tk	taka (Bangladesh)
TWh	terawatt-hour
UFG	unaccounted for gas

All dollar amounts are U.S. dollars unless otherwise indicated.

Overview

In the summer of 2012, India suffered the largest electrical blackout in history. Almost 700 million people—roughly equivalent to the entire population of Europe—lost power for two days. The power failure started when three of the country's five state-owned electricity grids failed. First to fail, on July 30, was India's northern grid. Revived after 14 hours of repair, it collapsed again the next day, quickly followed by the eastern and then the northeastern grids. The blackout stretched across roughly 2,000 miles, from India's western border with Pakistan to its eastern border with Myanmar. Trains were stranded on tracks; miners were trapped underground; traffic lights were extinguished, causing havoc on the roads; and millions of people were left without electric fans or air conditioners during the scorching heat of summer.

This power failure epitomizes the vulnerability of India's electricity sector. But India is not alone in struggling to keep the lights on. According to the most recent business surveys, conducted in 2011–15, South Asia had more frequent power outages than any other world region (Figure 1). Many of its countries rely on scheduled blackouts ("load shedding") to cope with the systemic shortages that occur as the supply of electricity continually falls short of the rapidly increasing demand. Firms reported almost daily blackouts, typically lasting more than five hours. Households had it even worse, reporting daily outages up to 10 hours in Bangladesh and up to 20 in some parts of Pakistan before 2014. The 2018 *Global Competitiveness Report*, which ranks 137 economies on the reliability of electricity supply, places Bangladesh at 101th, India at 80th, and Pakistan at 115th (Schwab 2018).

But power cuts are not the only concern. A bigger challenge is the large number of people forced to live without electricity 24/7. Among world regions, South Asia has the second-largest population living off the grid—255 million people in 2016, more than a quarter of all the people in the world living without access to electricity. Only Sub-Saharan Africa has more people not connected to the grid. As a result of low access rates and the low quality of supply, per capita electricity consumption in South Asia is

FIGURE 1 South Asia has the most unreliable power supply in the world

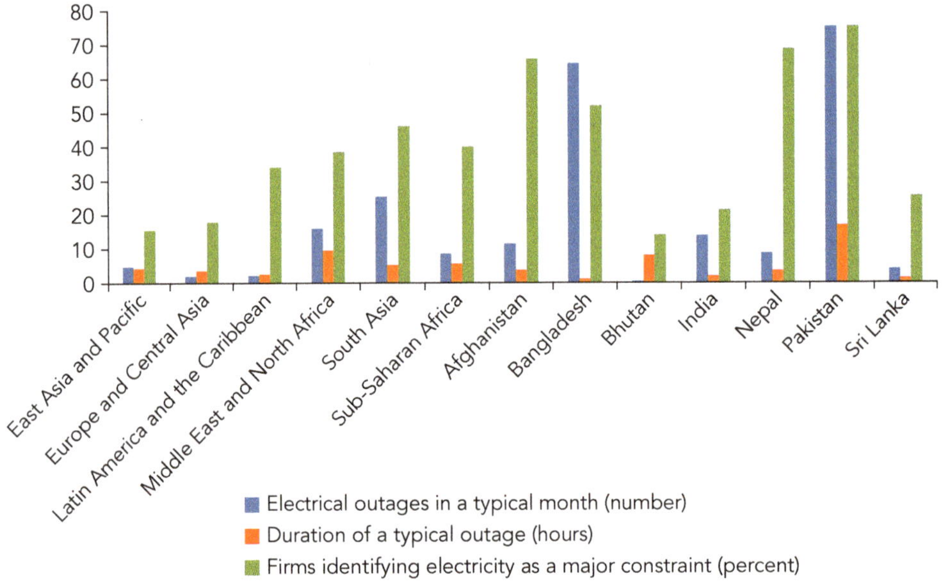

■ Electrical outages in a typical month (number)
■ Duration of a typical outage (hours)
■ Firms identifying electricity as a major constraint (percent)

Source: World Bank Enterprise Surveys in Afghanistan (2014), Bangladesh (2013), Bhutan (2015), India (2014), Nepal (2013), Pakistan (2013), and Sri Lanka (2011).

the second-lowest in the world (after Sub-Saharan Africa). At 707 kilowatt hours (kWh) a year in 2014, it is less than a quarter of the world average (Figure 2).

Inadequate access to electricity has important implications for economic development. In responding to World Bank Enterprise Surveys, almost half of business managers in South Asia identified lack of reliable electricity as a major constraint to their firm's operation and growth (see Figure 1). Indeed, they ranked blackouts as far more important than other barriers, including regulations and taxes, corruption, and human capital. Frequent blackouts force businesses to rely on generators, which produce electricity at a much higher cost than the grid. They force households to rely on kerosene lamps, a dirtier and costlier source of light. Lack of reliable electricity is also a major barrier to the economic advancement of underserved households, adversely affecting income, health, children's educational attainment, and gender equality (Samad and Zhang 2016, 2017, 2018).

Conventional wisdom suggests that inadequate investment in power infrastructure is the main cause of power shortages in South Asia. But a closer look at the data reveals a different picture. Over the decade ending in 2016, Bangladesh and India more than doubled their power-generation capacity, with average annual growth in capacity outstripping annual growth in gross domestic product (GDP). But in Bangladesh less than 80 percent of available capacity was operational most of the time (BPDB 2015, 2016);

FIGURE 2 South Asia has a quarter of the world's people without electricity—and the world's second-lowest regional per capita electricity consumption

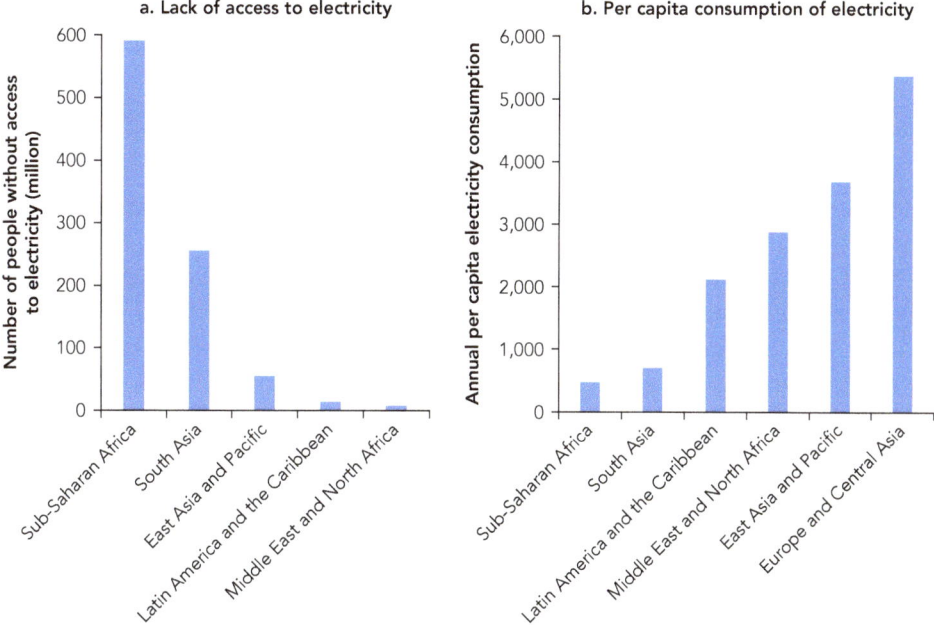

a. Lack of access to electricity

b. Per capita consumption of electricity

Source: World Development Indicators database.
Note: Data on access to electricity are for 2016. Data on per capita consumption are for 2014.

in India power shortages reached 5 percent of estimated demand in 2014, but up to 15 percent of coal power plants were left idle (CERC 2015). In fiscal 2018, when total installed capacity was more than twice the amount of peak demand, peak demand shortage still registered 2.1 percent in India (CEA 2018). Even in Pakistan, where capacity growth lags GDP growth, only 80 percent of available capacity was operational in fiscal 2014 (World Bank 2015a). Losses in transmission and distribution add to the shortages: India and Pakistan lose about a quarter of electricity in the network for both technical and commercial reasons, well above the 10 percent international norm.

South Asia thus faces an efficiency gap. Inefficiencies originating in every link of the electricity supply chain have resulted in upstream fuel shortages, poorly performing state utilities, and wasteful consumption downstream. Although there are multiple inefficiencies, most are attributable to three types of distortions: institutional distortions caused by state ownership and weak governance; regulatory distortions resulting from price regulation, subsidies, and cross-subsidies; and social distortions related to the negative externalities (such as emissions and associated health damage) of energy production and consumption.

Using microeconomic data from utilities, households, and firms, this report quantifies the economic cost of each type of distortion at each stage of power supply. The results show that the overall economic cost of distortions—ranging from 4 to 7 percent of GDP—is far greater than previously thought on the basis of analysis considering only the fiscal implications of distortions. Going beyond the traditionally defined power sector (generation, transmission, and distribution), the report suggests that some of the costliest distortions occur in upstream fuel supply and downstream access and reliability.

The report focuses on South Asia's three largest economies. Bangladesh, India, and Pakistan have a combined population of 1.6 billion, including almost 300 million people living in extreme poverty (subsisting on less than $1.90 a day) and 245 million lacking access to electricity. The three countries account for 98 percent of South Asia's electricity supply.

South Asia has made impressive progress in promoting the development of renewable energy in recent years. Bangladesh is a hotspot of the global off-grid solar power market. India now ranks fourth in the world in terms of installed wind energy capacity and sixth in solar-based capacity (Press Information Bureau 2017). Fossil fuel still plays a dominant role in power generation in South Asia, however. Bangladesh, India, and Pakistan together emitted 1.15 billion tons of carbon dioxide for power generation in 2015, almost as much as the power sectors of all the Organisation for Economic Co-operation and Development (OECD) countries in Europe. Both Bangladesh and Pakistan have also set ambitious targets for expanding the use of coal.

As a development imperative, improving the supply of electricity should be a first-order concern. But considering the role of power sector distortions is also important. Comprehensive sector reform that addresses inefficiencies at different stages of power supply could not only play a big part in increasing the supply of electricity but also in limiting the reliance on fossil fuel.

What This Study Adds

Many studies have examined the cost of power sector distortions in South Asia. They typically consider a narrow definition of the power sector—one that includes generation, transmission, and distribution and often omits upstream fuel supply and downstream access to electricity and reliability of supply. Most studies also focus on fiscal costs, ignoring the fact that, although there is no fiscal cost to a rural household lacking access to electricity or the atmosphere being polluted by coal-fueled generating plants, the economic costs are huge.

This report introduces two innovations. First, it goes beyond fiscal costs, evaluating the impact of distortions from a welfare perspective by measuring the economic cost of distortions through their impact on consumer wellbeing, producer surplus, and environmental costs. Second, it adopts a broader definition of the sector, one that covers the entire supply chain of power supply, including upstream fuel supply and downstream access and reliability (Figure 3).

Using a common analytical framework and covering all stages of power supply, this report provides what we believe to be the most comprehensive analysis to date of how

FIGURE 3 The report analyzes power sector distortions along the entire supply
chain of electricity

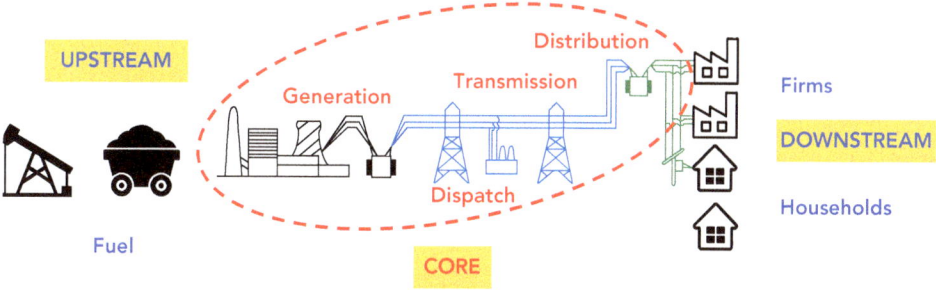

Source: Schematic of the core sector is from United States Department of Energy. Icons outside the core defined
by the dashed line are from the Noun Project, by the following artists: Oil well by Jason Dilworth, coal wagon by
Georgiana Ionescu, factory by pictohaven, and house by Adrien Coquet.

policy-induced distortions and externalities have affected social welfare in Bangladesh,
India, and Pakistan.

BEYOND FISCAL COSTS

Subsidies are often recognized as the main distortion in the power sector. Most studies
emphasize their fiscal implications. For example, on the basis of the difference between
regulated and market prices, the International Energy Agency estimates that subsidies in
India's power sector amount to 0.36 percent of GDP (IEA 2013). Accounting for direct
budgetary support by the government, the OECD estimates that subsidies in India's coal
sector represent less than 0.001 percent of GDP (OECD 2015). Although subsidies create
fiscal burdens, they also have redistributive effects. But, more important, subsidies contribute to energy shortages by distorting consumption and production and undermining
the performance of utilities. This report argues that the correct measure of the economic
cost of subsidies is thus not the fiscal costs but the loss in net output and consumer welfare.

Going beyond subsidies, the report also considers costs stemming from institutional
and social distortions: efficiency losses caused by state ownership and weak governance,
welfare losses resulting from lack of reliable access to electricity, and external (health
and environmental) costs from excessive fossil fuel–based energy production and consumption. Institutional and social distortions do not result in direct fiscal costs, but
they lead to economic losses that are often much larger than the losses from subsidies,
because the efficiency losses from high production costs, poor service quality, and environmental and health damage lead to first-order efficiency losses whereas pricing inefficiencies are likely to be second-order effects (Joskow 2008).

The study uses microeconomic data to estimate key parameters in each country.
It then uses these parameters to estimate the cost of institutional, regulatory, and social
distortions. The results suggest that the costs of institutional and social distortions are
several orders of magnitude higher than the fiscal costs of distortions.

BEYOND THE CORE: UPSTREAM AND DOWNSTREAM

Going beyond the core electricity sector of generation, transmission, and distribution, the study covers issues upstream (coal and gas) and downstream (households and firms). Inefficiencies upstream and downstream often contribute most to the total cost of distortions.

Fuel supply is a crucial part of power generation. According to plant-level data, at even highly subsidized prices, fuel costs represent roughly 47 percent of the short-run marginal costs of gas power plants in Bangladesh and 63 percent in Pakistan, and they account for 15–64 percent of the variable costs of coal power plants in India (CEA 2004, 2015). Shortfalls in coal and gas have led to idled generation capacity and increased the need for more expensive and/or dirtier alternative fuel. Upstream inefficiencies can therefore quickly trickle down to consumers in the form of power cuts, costly electricity, and pollution. As this report shows, social distortions from coal use in India and the underpricing and inefficient allocation of gas in Bangladesh and Pakistan are among the largest sources of the overall economic cost of power sector distortions in those countries.

For the downstream population, power shortages represent a barrier to social and economic development. Lack of reliable access to electricity is associated with lower income, higher poverty, poorer health and education, and less gender equality (Figure 4).

FIGURE 4 Access to electricity is associated with higher income and better social outcomes in Bangladesh, India, and Pakistan—and the results are much stronger if the electricity is reliable

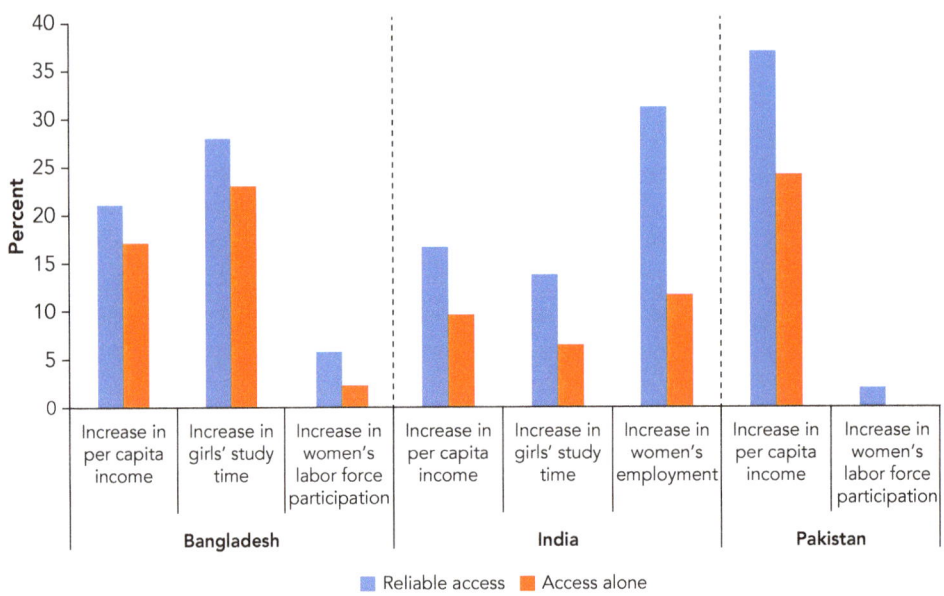

Source: Estimation based on household surveys in Bangladesh, India, and Pakistan. See also Samad and Zhang (2016, 2017, 2018).
Note: The effects of electrification on girls' study time and the effects of power outages on women's labor force participation in Pakistan are not estimated because data are not available.

Unreliable power supply also adversely affects the operation and growth of firms. Large businesses try to cope by investing in captive generators; small and medium-size businesses are usually unable to do so (Grainger and Zhang 2017). In manufacturing and services combined, the total losses in annual output attributable to power shortages amounted to $1.1 billion and $22.7 billion in Bangladesh and India, respectively, in fiscal 2016, and $8.4 billion in Pakistan in fiscal 2015.

Massive Electricity Shortages

Over the past few decades, countries in South Asia have substantially expanded electricity supply, improved access, and promoted market-oriented reforms. The combined generation capacity in Bangladesh, India, and Pakistan grew from 198 gigawatts (GW) in 2007 to 376 GW in 2017; the share of households with access to electricity in the three countries rose from less than 70 percent in 2007 to an officially estimated 86 percent in 2016, according to the latest World Development Indicators. All three countries have launched power sector reforms to encourage private investment since the 1990s.

Despite this progress, South Asia continues to face electricity shortages in terms of both access and quality of supply. On a per capita basis, total installed electricity-generating capacity still falls behind the world average: One-quarter of the world's people live in South Asia, but the region has just 5 percent of global electricity-generating capacity. In addition, heavy reliance on fossil fuel for power generation poses a daunting challenge as the region struggles to balance the need for energy with its environmental consequences.

LOW ACCESS AND LOW QUALITY OF SUPPLY

South Asia has the world's second-lowest rate of access to electricity, after Sub-Saharan Africa. The 255 million people in South Asia who lack an electricity connection represent roughly 14 percent of the region's population.

The lowest access rate in South Asia is in Bangladesh, where 24 percent of the population lived off the grid in 2016 (31 percent in rural areas) (Table 1). India achieved 100 percent village electrification in 2018. But at the household level, its rural access rate, at 81 percent in 2017, is still the third-lowest in South Asia. In Pakistan 99 percent of the population has access to grid electricity, according to official statistics, but estimates based on census data and the number of connections reported by utilities suggest that access to grid electricity was only about 74 percent in 2016 (IEA 2017). A household survey carried out by the International Finance Corporation in 2014 even suggests that up to 35 percent of the population in Pakistan may still live off the grid (IFC 2015).

For people nominally connected to the grid, access to electricity can be uneven and unreliable, characterized by frequent, long-lasting power outages. Outages often occur because of technical failures. In South Asia they also reflect the efforts of utilities to

TABLE 1 South Asia has low rates of access to electricity, especially in rural areas

Country	Total (percent of population)	Rural areas (percent of population)
Afghanistan	84.1	79.0
Bangladesh	75.9	68.9
Bhutan	100.0	100.0
India	86.1	81.0
Maldives	100.0	100.0
Nepal	90.7	85.2
Pakistan	99.1	98.8
Sri Lanka	95.6	94.6

Source: World Development Indicators database. Indian data are from the Indian rural electrification program's (Saubhagya) dashboard, updated as of October 2017.
Note: Data are for 2016 except for India.

cope with power shortages through scheduled power cuts (load shedding). In World Bank Enterprise Surveys conducted in 2011–15, business managers in the region reported that power cuts occur almost every day, with an average duration of 5.3 hours. By comparison, managers in East Asia reported one outage every nine days, and managers in Sub-Saharan Africa reported one outage every four days. To deal with power disruptions, almost half of firms in South Asia own or share a generator.

Within South Asia, Bangladesh and Pakistan had the most severe power shortages. In both countries, electricity demand routinely exceeded supply, triggering crippling blackouts nationwide.

The officially reported power shortages in Bangladesh, India, and Pakistan have all declined in recent years, thanks to new capacity addition, the decline in global oil price until recently, and, in India, lower than expected growth in demand. But these official figures almost certainly underestimate the true power deficit: because electricity demand is often defined as the amount of electricity distribution utilities buy, it does not account for demand by people who remain unserved or underserved. Lack of reliable access to electricity stymies the growth of businesses and disrupt people's daily lives, periodically prompting protests that sometimes turn violent (The Guardian 2012).

DIRE ENVIRONMENTAL AND HEALTH CONCERNS

As South Asia has expanded its electricity supply, the region has become increasingly dependent on fossil fuel for both grid electricity and captive power generation (Figure 5). This dependence has helped create some of the most polluted cities in the world. Fossil fuel–based power generation is the largest source of carbon dioxide emissions in the region.

FIGURE 5 South Asia has become increasingly dependent on fossil fuel for power generation

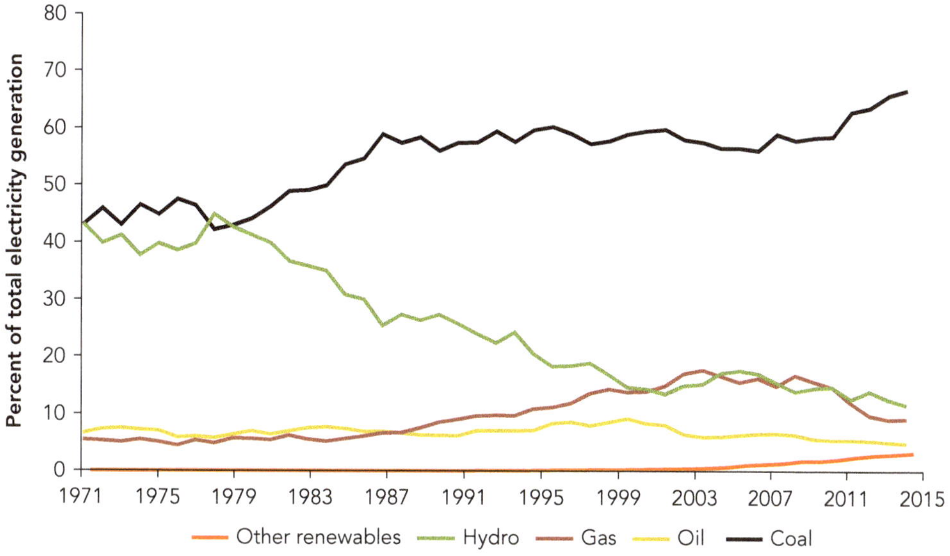

Source: World Energy Statistics and Balances database (IEA 2018).
Note: Figure shows the fuel mix for both grid electricity and captive power generation.

Burning coal and diesel also releases numerous toxic pollutants. The most harmful is fine particulate matter 2.5 micrometers in diameter or smaller, known as $PM_{2.5}$. These particles, less than 1/30th the width of a human hair, can be inhaled deep into the lungs, causing illness and premature death.

The population in South Asia is exposed to some of the world's highest combustion-related concentrations of $PM_{2.5}$ (Health Effects Institute 2017). At 89 micrograms per cubic meter in Bangladesh, 74 in India, and 65 in Pakistan, the annual population-weighted average concentrations are many times the World Health Organization's safe limit of 10 micrograms per cubic meter. The trend is also worrisome: Between 2010 and 2015, Bangladesh and India experienced the steepest increases in $PM_{2.5}$ concentration among the world's 10 most populous countries.

With worsening air quality, the three countries also have some of the highest mortality rates attributable to ambient air pollution (Map 1). Between 1990 and 2015, the annual number of deaths attributable to $PM_{2.5}$ exposure increased by 64 percent in Pakistan, 51 percent in Bangladesh, and 48 percent in India. In the three countries combined, the annual number of deaths attributable to $PM_{2.5}$ rose by 50 percent over the period, from 900,900 in 1990 to 1,347,900 in 2015 (Health Effects Institute 2017).

MAP 1 South Asia has some of the world's highest mortality rates associated with exposure to fine particulate matter

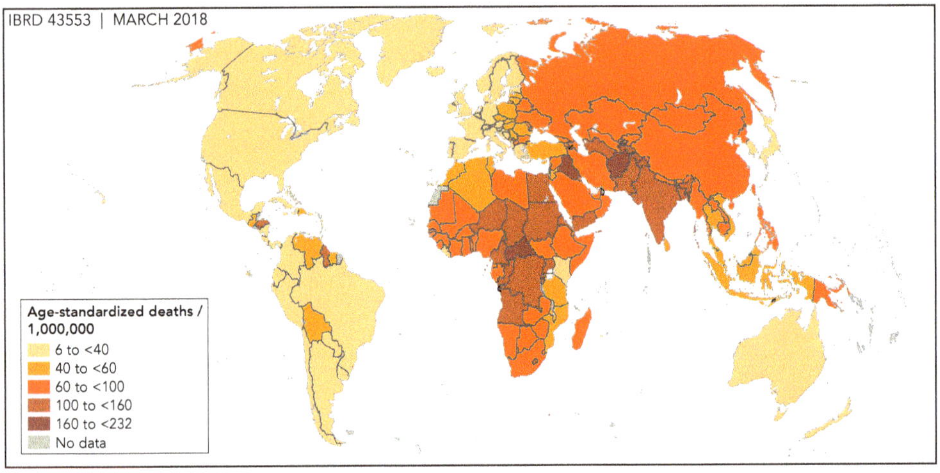

Source: Health Effects Institute 2017.

Three Types of Distortions

Multiple distortions have contributed to the power crisis in South Asia. They can be grouped into three categories.

INSTITUTIONAL: NO MARKET

Institutional distortions in the energy sector stem from the dominance of government ownership, the lack of competition, and soft budget constraints, under which governments have repeatedly bailed out heavily indebted utilities. Despite recent reforms, state-owned enterprises continue to dominate the sector. Government planners, not the market, allocate fuel supplies and set prices. Because the market plays a limited role in penalizing underperformance and rewarding efficiency, energy suppliers, especially public ones, face little pressure to control costs and maximize outputs.

The inefficiency of state-owned enterprises is exemplified by their performance in power generation. Using multiyear data at the level of thermal power plants, this report finds an astonishingly wide gap in efficiency between public and private plants (Figure 6). The conclusion holds even after controlling for differences in the age, capacity, location, technological, and operational characteristics of power plants. All else equal, a public plant uses substantially more fuel than a private one to produce the same amount of electricity—on average, up to 29 percent more in Bangladesh, 16 percent more in India, and 20 percent more in Pakistan. Some of this difference may be explained by the type of power purchase agreements signed by private plants, which allows them to be

FIGURE 6 Public power plants are substantially less efficient than private
ones: Bangladesh and Pakistan as examples

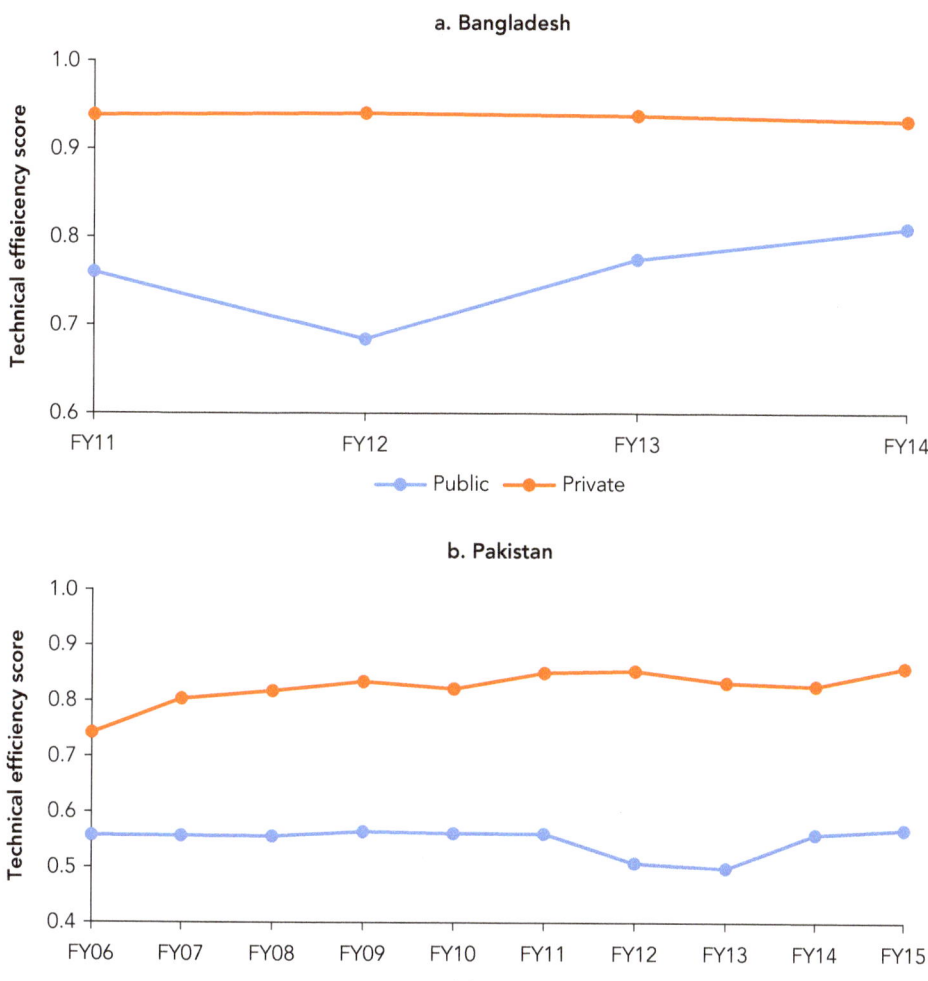

Source: Estimation based on plant-level data from Bangladesh Power Development Board (BPDB) annual reports
(2011–14) and the Pakistan National Electric Power Regulatory Authority's (NEPRA) State of Industry Report (2006–15).
Note: Technical efficiency score measures the ratio of actual output to maximum feasible output. Private plants
refer to independent power producers but not rental power plants in Bangladesh. FY = fiscal year.

dispatched at optimal load factors. But the efficiency gap could also reflect differences
in managerial behavior across ownership types.

The inefficiency in generation imposes substantial opportunity costs, especially
given the coal and gas shortages in all three countries. Simulation analysis shows that
if public power plants eliminated their operational inefficiency, Bangladesh and India
could reduce about 50 percent and Pakistan roughly 25 percent of their unserved energy
demand with no new investment in generation capacity (Figure 7).

FIGURE 7 Institutional distortions in power generation exacerbate electricity shortages: India as an example

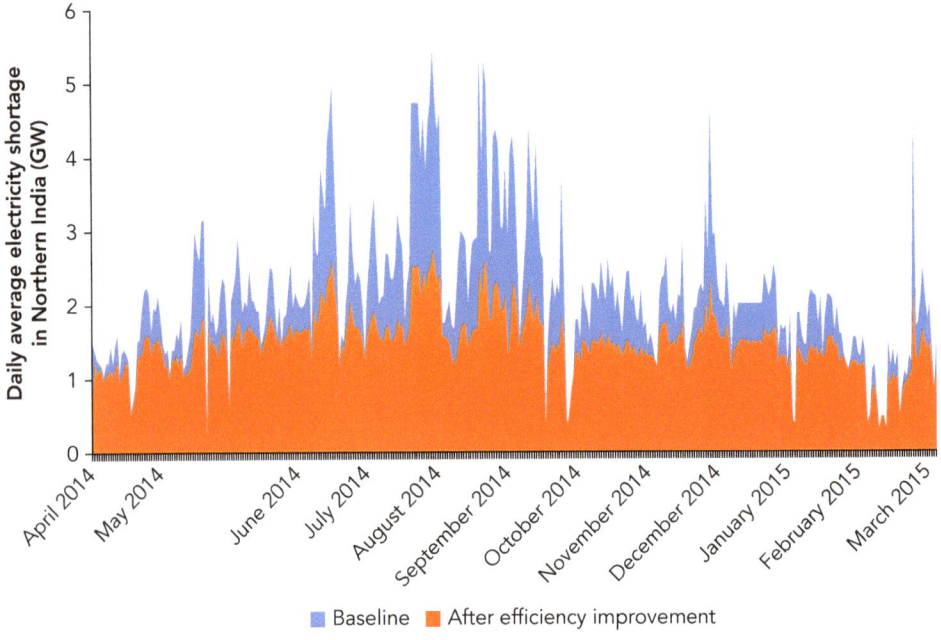

Source: Simulation based on Indian Central Electricity Authority (2000–12) and daily reports by the Northern Regional Load Dispatch Center.
Note: GW = gigawatt.

Soft budget constraints often exacerbate the inefficiency inherent in state ownership, as most evident in the power distribution sector. In India and Pakistan, hefty losses of electricity in distribution, along with poor recovery of overdue electricity bills, have given rise to alarming levels of debt in the sector and prompted repeated government bailouts. India's central government launched rescue operations to bail out loss-making distribution companies three times since fiscal 2001. In Pakistan the government periodically pays down the "circular debt" resulting from the combined losses in transmission and distribution—a debt that reached a staggering $9 billion by the end of fiscal 2012 (USAID 2013). These government rescues have not helped eliminate debt or electricity losses over the long term (Figure 8).

Institutional distortions also reduce allocative efficiency. In India the allocation of coal blocks (leases) favors government-owned power utilities. In Pakistan natural gas is routinely diverted from power generation to other sectors, even though gas is estimated to have the greatest economic benefit in the medium term when used in power generation (USAID 2011). In Bangladesh not only do less efficient power plants receive privileged access to gas (Figure 9) but they also are often brought into production before other generators, despite being two to three times as costly to operate (World Bank 2015b). Inefficient allocation of inputs and outputs in the electricity sector exacerbates power shortages. In Bangladesh and Pakistan,

FIGURE 8 Distribution utilities in India and Pakistan incur high electricity losses

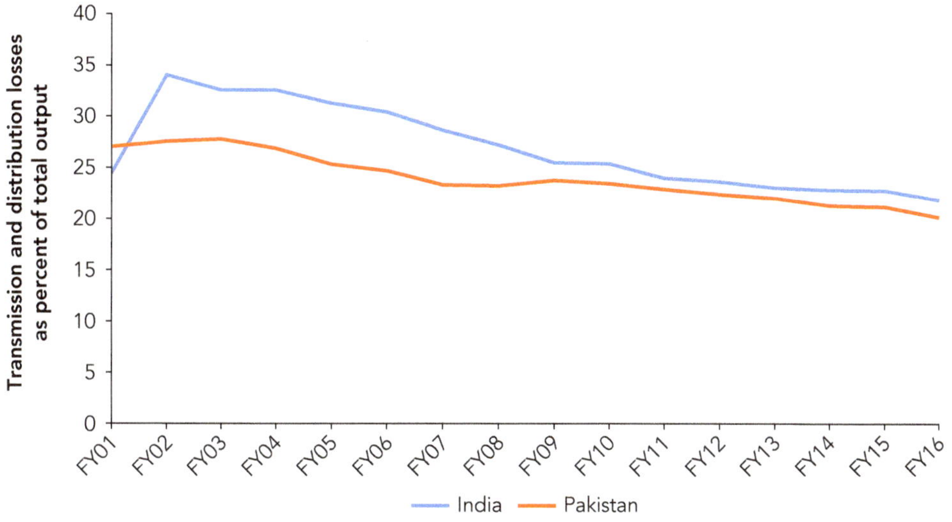

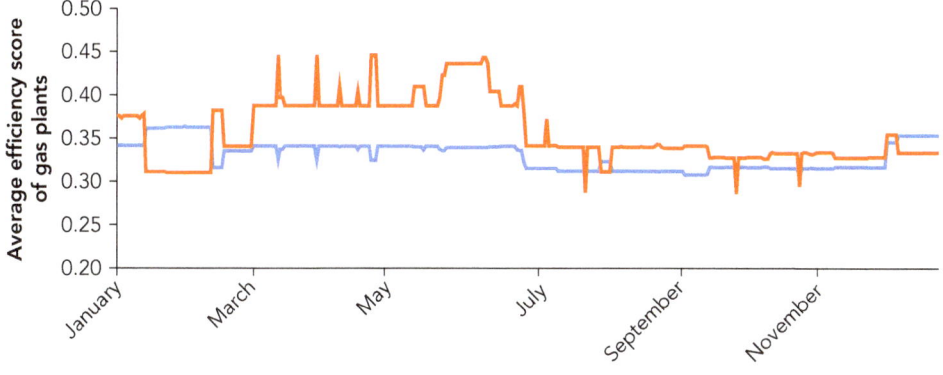

Source: Indian Central Electricity Authority (2017) and Pakistan National Transmission and Dispatch Company (2016).
Note: FY = fiscal year.

FIGURE 9 Less efficient power plants receive privileged access to gas in Bangladesh

Source: Based on daily reports by the Bangladesh Power Development Board (January 1–December 31, 2014).
Note: Efficiency score is the ratio of electricity output to gas input in calorific value. Average efficiency is weighted by capacity.

it also increases the need for oil-based power generation, contributing to heavier emissions.

REGULATORY: MARKET BUT DISTORTED

Regulatory distortions arise from subsidies and the mispricing of coal, gas, and electricity. Energy subsidies are widespread in South Asia. In addition to creating fiscal burdens, they distort incentives for production and consumption and undermine the performance of utilities.

In the upstream fuel sector, coal and gas are priced substantially below their opportunity cost, even without factoring in their external costs to the environment. In Bangladesh the international benchmark price of natural gas is almost 11 times the domestic price for power generation. In India the price of coal for the power sector (along with the fertilizer and defense sectors) was 17 percent lower than the price charged to other sectors (CIL 2018); it was a third lower than the spot market price. Pakistan has a two-tier gas market. Imported liquified natural gas (LNG) is broadly charged at the full cost to consumers, but domestic gas was priced at roughly 36 percent of the international benchmark in fiscal 2016 (Figure 10).

Underpricing coal and gas contributes to fuel shortages, not only because it encourages wasteful energy consumption but also because it reduces suppliers' interest in upstream exploration and production. In Bangladesh and Pakistan, several large gas development projects have been abandoned because of the government's unwillingness to raise tariffs to allow cost recovery with reasonable returns. Because of the dependence on coal or gas for power generation, upstream fuel shortfalls have quickly cascaded into idled capacity downstream. Fuel shortages left an average 10 percent of gas capacity in Bangladesh and 15 percent of coal capacity in India stranded in 2014. In Pakistan shortages of gas for power generation were made up through expensive imported oil, increasing both electricity costs and trade bills.

FIGURE 10 The price of domestic natural gas is much lower than the international price in Bangladesh and Pakistan

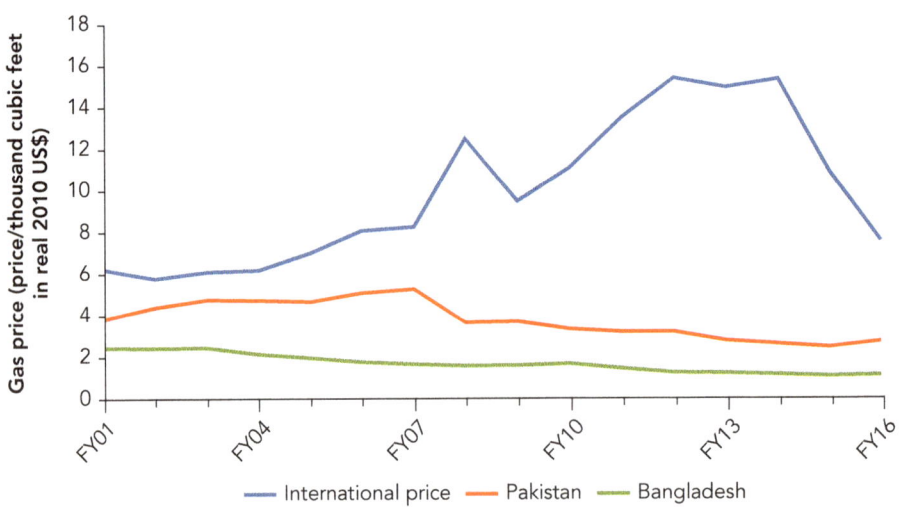

Source: Petrobangla Annual Report (2016); Hydrocarbon Development Institute of Pakistan (2001, 2007, 2014, 2016); Pakistan Ministry of Finance (2017); and the World Bank Global Economic Monitor Commodities database.
Note: FY = fiscal year.

The core electricity sector also underprices. For households and farmers, electricity is priced lower than the cost for utilities to buy it—25 percent lower in Bangladesh, and 22 percent lower in India in fiscal 2016, and 7 percent lower in Pakistan in fiscal 2015 (Figure 11). In addition, irregularities in billing and rampant theft of electricity constitute a de facto implicit subsidy. Distorted tariffs combined with unpaid subsidies have contributed to the deteriorating financial situation of distribution utilities. It not only compromises investment and maintenance (Pargal and Banerjee 2014) but also creates perverse incentives for utilities to underserve loss-making customers, especially in rural areas, where the cost of service is high.

In India, for example, analysis for this report using nighttime satellite images for 2013 shows that areas adjacent to newly electrified villages subsequently experienced worse power outages after the villages were connected to the grid. As more low-paying consumers joined the grid, distribution utilities may have been either unable or unwilling to invest in maintaining and upgrading infrastructure to expand the power supply.

Regulatory distortions also take the form of cross-subsidies between consumer groups. In the Indian rail system, for example, coal freight cross-subsidizes passenger service. This cross-subsidization leads to higher electricity prices for consumers and undermines efficiency and investment in freight rail. The resulting constraints in rail capacity have created bottlenecks in coal supply in India. Econometric analysis shows that every 1 percent increase in distance between coal mines and the power plants they serve increases the plants' coal shortage by 14 percent, reduces their utilization rate by 3 percentage points, and increases their output shortage by 10 percent on

FIGURE 11 Electricity tariffs in India illustrate the extent to which residential and agricultural consumers are subsidized

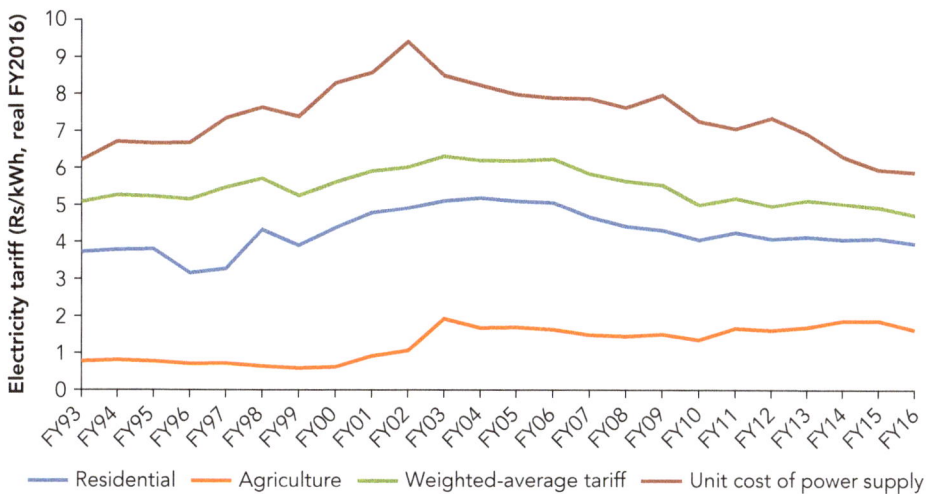

Source: Indian Power Finance Corporation (2003, 2007, 2017) and Indian Planning Commission (2000, 2001, 2002, 2012, 2014, 2015).
Note: Unit costs of power supply for fiscal 2015 and 2016 are based on annual plan projection. Domestic prices for fiscal 2002 and 2003 are interpolated because the original data files did not include them. FY = fiscal year; Rs/kWh = rupees per kilowatt-hour.

average (Figure 12). An additional 34 million tons of coal could be delivered each year if railway distortions were removed and coal shortages were no longer linked to the distance to coal mines.

In the core sector, industrial and commercial users of electricity are often overcharged to compensate for the lower rates for households and farmers (Figure 11). Although the higher electricity prices for these consumers help relieve the fiscal burden on the government, they lead to unintended consequences downstream. Because electricity is required as a primary input in nearly every sector, overcharging industrial and commercial consumers raises the prices of almost all goods and services. Meanwhile, high electricity tariffs for industry undermine export competitiveness, especially for energy-intensive producers (Figure 13). Removing the cross-subsidies could increase India's net manufacturing exports by 1–3 percent depending on the sector (Figure 13).

FIGURE 12 Distance to coal mines is correlated with worse coal shortages and lower power generation in India

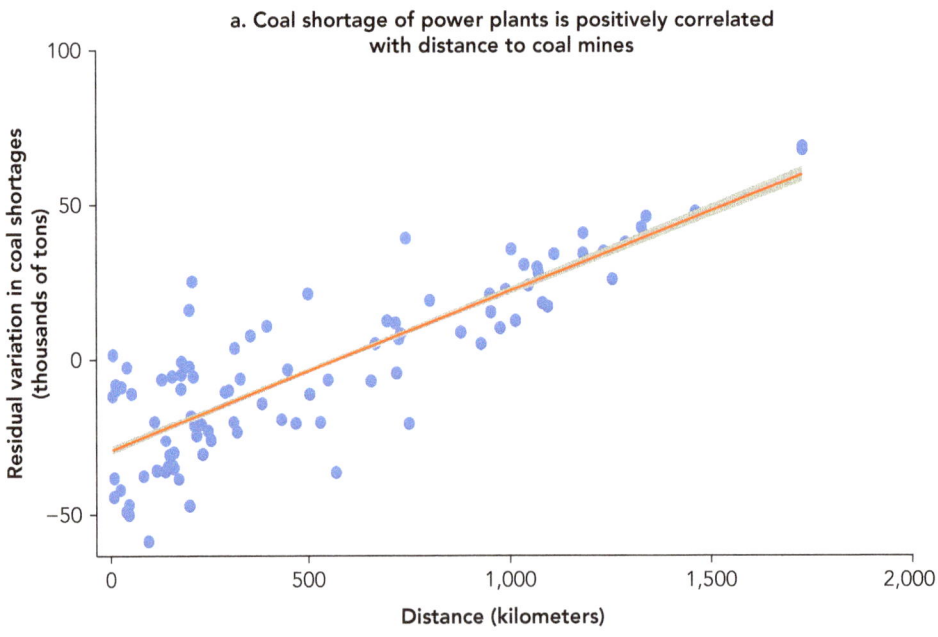

a. Coal shortage of power plants is positively correlated with distance to coal mines

figure continues next page

FIGURE 12 Distance to coal mines is correlated with worse coal shortages and lower power generation in India *(continued)*

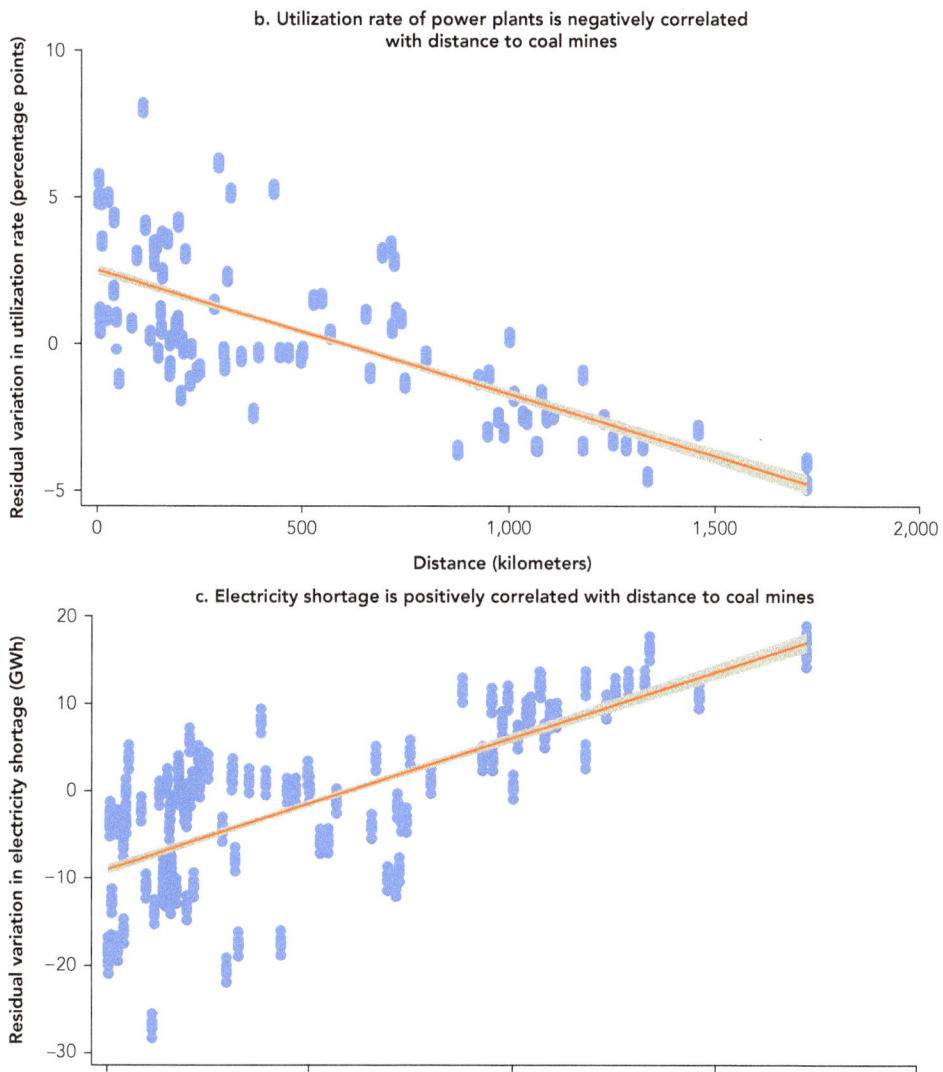

b. Utilization rate of power plants is negatively correlated with distance to coal mines

c. Electricity shortage is positively correlated with distance to coal mines

Source: Coal linkage data are from the India National Thermal Power Corporation and the Central Board of Irrigation and Power. Daily actual and normative required coal stock data are from the Central Electricity Authority of India (2008–16). Data on monthly power generation of coal plants are from the Central Electricity Authority of India (2012–16).
Note: Coal shortages are daily average shortages, defined as the normative coal stock minus the actual coal stock. Electricity shortage is defined as a plant's targeted output minus its actual output. The vertical axis is the difference in residuals from regressions with and without controlling for distance between power plants and coal mines. Other independent variables in the regression include capacity, age, age squared, quality of coal, year, month, and region fixed effects. Gray shaded areas are 95 percent confidence intervals. See chapter 4 for details about the regression analysis. GWh = gigawatt-hour.

FIGURE 13 Cross-subsidies in electricity tariffs undermine the competitiveness of Indian industries

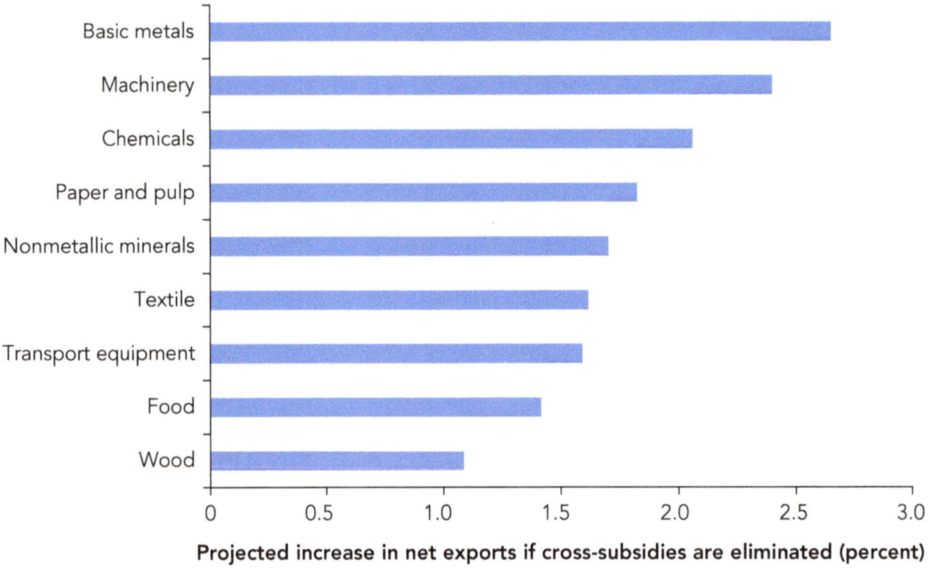

Projected increase in net exports if cross-subsidies are eliminated (percent)

Source: Chan, Manderson, and Zhang 2017.

SOCIAL: MARKET BUT WITH EXTERNALITIES

Social distortions reflect the negative externalities of energy production and consumption, including the health costs of coal mining and combustion and the climate change effects from burning fossil fuel. In addition, in India the provision of heavily subsidized electricity to farmers for pumping water has encouraged water-intensive farming practices and triggered the depletion of groundwater.

Fossil fuels dominate the fuel mix for power generation in South Asia. In 2015 gas accounted for 81 percent of electricity generation in Bangladesh and coal for 75 percent in India. In Pakistan oil accounted for 37 percent and gas for 27 percent. In addition to contributing to climate change, emissions from fossil fuel–based power generation have well-documented adverse effects on health. In India the air pollution produced by coal-fired power plants is a leading risk factor for death, contributing to the loss of about 2.3 million years of healthy life (disability-adjusted life years) in 2015 (Global Burden of Disease MAPs Working Group 2018). Although gas is cleaner than coal, its combustion produces nitrogen oxides—precursors to ground-level ozone (urban smog) that can cause various respiratory diseases.

When pricing fails to account for these external costs of fossil fuel consumption, emissions are excessive. Imposing an environmental tax on emissions can be a cost-effective way to reduce air pollution; it could also pave the way for a forceful turn toward the development of renewable energy. India is among the few countries that have introduced an environmental tax on coal consumption. But its Clean Environment Cess offsets less than 3 percent of the marginal environmental and health damage caused by coal-based power generation. Bangladesh and Pakistan have no such environmental tax.

The net social benefit from achieving full-cost pricing can be approached as the sum of avoided environmental and health damage, increased revenue from environmental taxation, and forgone consumer and producer surplus. This annual benefit is estimated at $345 million in Bangladesh, and $35.4 billion in India.

Improving the efficiency of gas allocation and use is another way to reduce pollution in Bangladesh and Pakistan. Waste in gas consumption has led to greater reliance on furnace oil and diesel for power generation. These liquid fuels are not only more expensive but they also out-pollute gas by 30–600 percent, depending on the type of emissions (IPCC 2006). Simulation analysis shows that improving fuel efficiency and channeling gas from less efficient to more efficient uses would reduce the consumption of liquid fuel and cut annual carbon dioxide emissions by 250,000 tons in Bangladesh and 1.8 million tons in Pakistan (Figure 14).

FIGURE 14 Improving the operating efficiency of gas units would reduce the use of oil: Evidence from Bangladesh

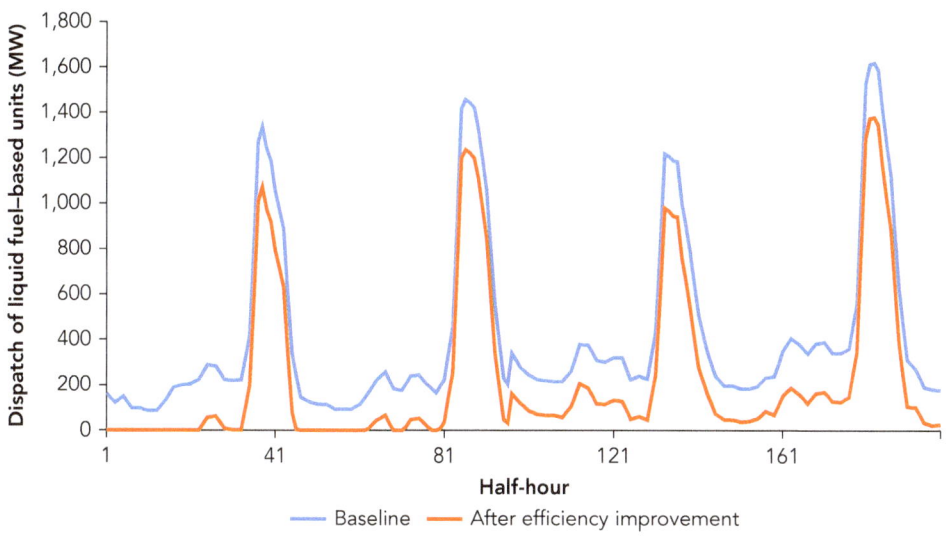

Source: Simulation based on daily reports by the Bangladesh Power Development Board, January 1–December 31, 2014.
Note: For illustration, this figure shows data for January 1–4, 2014. The pattern remains the same for the year as a whole. MW = megawatt.

Another social cost of power sector distortions comes from the heavy reliance on kerosene lighting and captive power generation in South Asia. Households and small businesses lacking reliable access to electricity turn to kerosene lamps to meet basic lighting needs, using an estimated 244 million lamps in the region (Tedsen 2013). Many studies report a strong association between kerosene lighting and tuberculosis risk and respiratory infections (WHO 2015). Analysis in this report shows that households without a connection to the grid consume 14–88 percent more kerosene than households with a connection, all else equal. In India access to electricity is associated with a 7.4 percent reduction in the number of days of illness. The health-related income loss from lack of access to electricity is estimated at at least $410 million a year (Samad and Zhang 2016).

Kerosene lamps also contribute to emissions of ambient black carbon, a major climate warmer in the atmosphere, second only to carbon dioxide. Black carbon remains in the atmosphere for only a few days, but during that time a single gram has several hundred times the global warming impact that the same amount of carbon dioxide has over 100 years (Jacobson and others 2013). Black carbon emissions also contribute to snow and ice melting in the Himalayas and increase the disruption of the South Asian monsoon patterns (Shindell and others 2012).

South Asia already experiences some of the greatest warming effects of black carbon emissions from residential kerosene lighting (Map 2). The annual environmental cost of black carbon emissions from kerosene lighting is estimated at $0.6 million in Bangladesh, $6.4 billion in India, and $2.1 million in Pakistan.

Another consequence of unreliable access to grid electricity is the increased use of fossil fuel–based captive generation, such as diesel generators. Captive generators are usually less efficient than utility-scale power plants. They are also located closer to population centers and at ground level (without high stacks of utility power plants). For all of these reasons, they are likely to have a greater environmental effect for a given amount of electricity produced.

Another social distortion stems from electricity subsidies for agriculture, which have contributed to the overexploitation of groundwater, particularly in India and parts of Pakistan (Figure 15). Electricity tariffs for the agricultural sector were estimated to be 70 percent lower than the average cost of electricity supply in India in fiscal 2016 (Indian Planning Commission 2015; Power Finance Corporation 2017).

Empirical evidence shows that farmers are price sensitive in their use of irrigation water (Veettil and others 2011). When the cost of water extraction is artificially low, farmers are less likely to adopt water-conserving irrigation technologies and more likely to shift to water-intensive crops such as rice. Many studies show a link between excessive agricultural electricity use and groundwater depletion (Badiani and Jessoe 2013).

Satellite images reveal a strikingly high rate of groundwater extraction in India. Groundwater extraction in Rajasthan, Punjab, and Haryana (including Delhi) in

MAP 2 Warming effects of black carbon emitted by kerosene lamps are greatest in South Asia

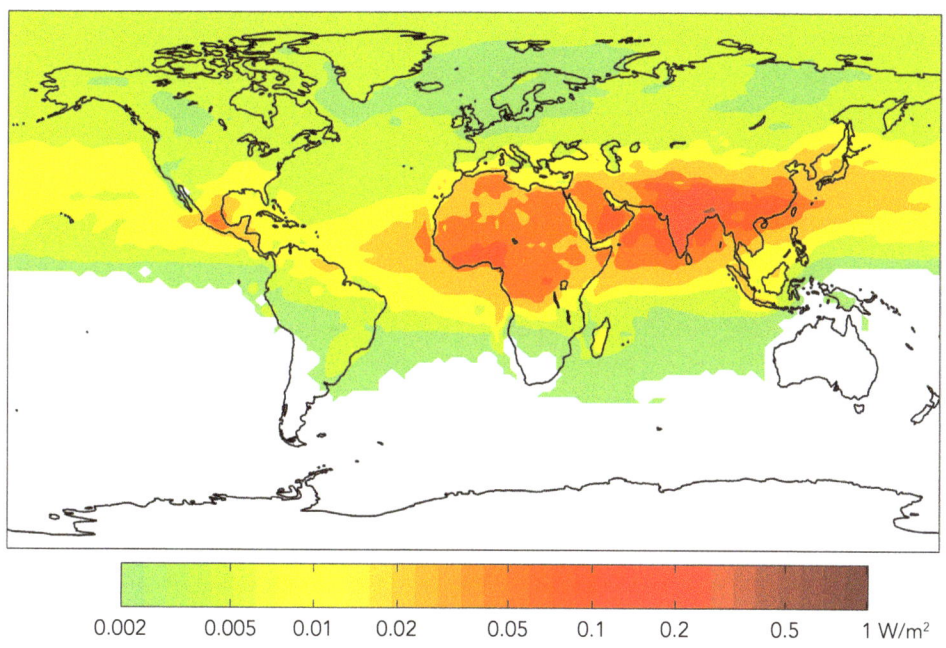

Source: Lam and others 2012.
Note: W/m² = watts per square meter.

FIGURE 15 Groundwater extraction has surged in India

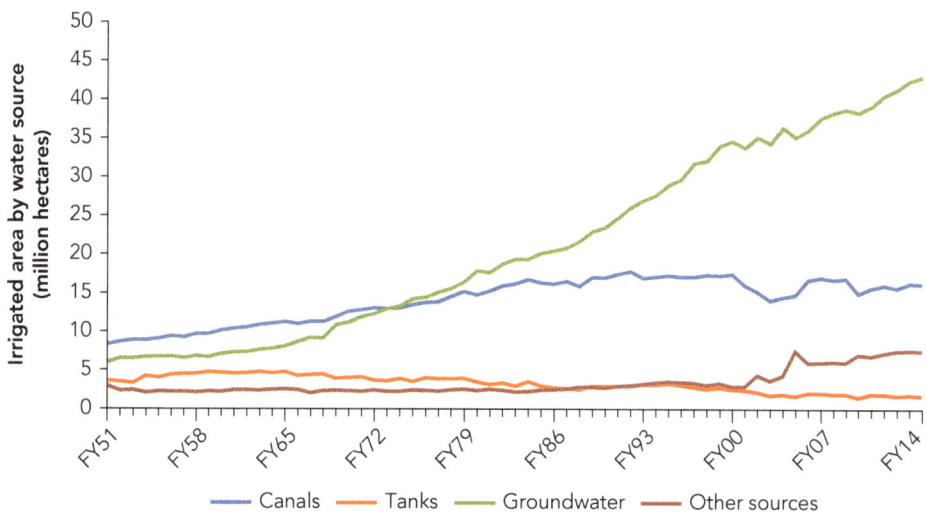

Source: Indian Ministry of Agriculture and Farmers Welfare, accessed through Indiastat.
Note: FY = fiscal year.

2002–08 was equivalent to a net loss of 109 cubic kilometers—twice the capacity of the country's largest surface-water reservoir (Rodell, Velicogna, and Famiglietti 2009). This rate of extraction is unsustainable. With about 60 percent of agriculture depending on groundwater for irrigation, and 85 percent of the rural population and 45 percent of the urban population relying on it for drinking water, the depletion of groundwater poses a significant risk to long-run food and water security in India (Sekhri 2013).

Conclusion

The full cost of distortions in the power sector is far greater than previously estimated on the basis of fiscal cost alone. Some of the largest costs are upstream or downstream, making the case for a stronger prioritization of power sector reform.

The total annual economic cost of power sector distortions is conservatively estimated at about $11.2 billion in Bangladesh (5.0 percent of GDP) in fiscal 2016, $86.1 billion (4.1 percent of GDP) in India in fiscal 2016, and $17.7 billion (6.5 percent of GDP) in Pakistan in fiscal 2015. In Bangladesh the underpricing of gas is the largest source of economic cost, responsible for an annual loss of $4.5 billion (2.0 percent of GDP). In India the environmental effects from excessive coal use are the largest source of cost, estimated at $35.4 billion a year (1.7 percent of GDP). In Pakistan the impact of the lack of reliable access to electricity on households and firms is the largest source, costing roughly $12.9 billion a year (4.8 percent of GDP).

These results suggest that the potential gains from power sector reform are huge. They include cost savings for utilities; income gains for households and firms; reductions in air pollution and health damage for the population; and lower subsidies to state-owned utilities, higher tax revenues, and lower public health spending for governments.

It is important to make power sector reform a top priority. Few other reforms could quickly yield economic gains of a similar magnitude. By expanding access to electricity and improving the quality of supply, power sector reform would also directly benefit poor households. A narrow focus on liberalizing the price of electricity should be avoided, however, because regulatory distortions in the core sector of electricity are often not the most important source of economic cost; and, in the absence of institutional reforms, inefficiencies of energy companies are passed onto consumers. The highest payoffs are likely to come from institutional reforms, the expansion of reliable access to electricity, and the appropriate pricing of carbon and emissions of local air pollutants.

POLICY RECOMMENDATIONS

The analysis points to several implications for the implementation of reforms.

Focus beyond the Core Sector

Achieving a reliable and sustainable electricity supply requires looking beyond the core power sector to address distortions in the upstream fuel sector. Doing so calls for measures to introduce effective competition in an otherwise monopolistic fuel market and to limit the government's political interference in operation and investment.

Pricing reform is also important. Fuel subsidies do not always have a large direct budgetary impact, but their opportunity cost is much greater than that of electricity subsidies in South Asia. Pricing that reflects the full economic cost of fuel would encourage production, curtail demand and emissions, and facilitate the efficient allocation of fuel across sectors. Diversifying the fuel portfolio to include different types and sources of fuel—by, for example, increasing regional energy cooperation and scaling up the development of previously untapped renewable resources—makes sense, because depending primarily on a single fuel raises reliability concerns.

Think beyond Investment

Although urgently needed in some segments of the power sector, investment alone is unlikely to solve the power crisis in South Asia. A big contributor to power shortages is inefficiency. Competition and private participation can improve operating efficiency. Competition can be promoted by ensuring nondiscriminatory access to fuel for public and private producers alike, by dispatching generation in merit order from lowest to highest cost, and by removing discriminatory charges on consumers buying electricity from the open market. In addition to outright privatization, other ways to tap private sector initiative include franchise arrangements in electricity distribution and contracts to outsource system operations and maintenance. In the absence of market competition, incentive-based regulation—such as price cap and yardstick competition mechanisms—can be used to reward more efficient operation. It is also important to prioritize investment to address electricity supply bottlenecks. With greater private sector participation and a more decentralized investment pattern, pricing mechanisms such as locational marginal pricing for transmission can provide signaling on where investment should be targeted.

Reform beyond Corporatization

Corporatization has been a key government strategy for power sector reform in South Asia. But, without fundamental changes in incentive structures, it is no guarantee of meaningful changes in performance. Because the government remains the controlling owner, corporatized utilities are still susceptible to political pressure. Moreover, the separation of management and control implies asymmetric information and agency costs. And, with or without corporatization, when firms believe that they will not be allowed to fail, they have little incentive to reduce losses. The effectiveness of corporatization thus depends on preventing inefficient political interference and soft budget constraints.

Floating newly corporatized companies on the stock market, which can play a unique role in monitoring and rewarding managerial efforts, has also been shown to help turn around firm performance.

Prioritize Quality, not Just Access

Achieving universal access to electricity brings a broad range of social and economic benefits and should remain high on governments' agenda. But merely ensuring connectivity is not enough. Unreliable supply of electricity discourages households and businesses from adopting electricity and limits the potential gains from electrification. As a result of regulatory and political imperfections, grid extension can undermine the quality of electricity service. Where electricity prices are too low to recover costs, adding new electricity connections inevitably puts greater strain on the grid because the system is forced to absorb more loss-making customers. Electoral incentives may create a bias favoring short-term, more visible investment in grid extension over long-term, hidden efforts in grid maintenance. In a budget-constrained environment, the drive toward quantity can come at the expense of quality for both existing and new customers.

To ensure the quality of electricity supply, it is important to remove electricity subsidies, so that utilities have the resources to invest in the long-term reliability of the grid. Cost-recovery tariffs also eliminate perverse incentives to underserve loss-making customers.

A powerful way to improve quality is to engage citizens in monitoring service delivery. Also critical is improving the collection and sharing of data on power outages. Understanding where and whose power gets cut improves accountability. Where utilities may underreport load shedding or resist sharing outage data, high-frequency satellite imagery of night lights data can provide an alternative means of monitoring power supply disruptions in close to real time.

Accompany Reforms with Compensation

Energy price reform requires large price increases. But price hikes can cause immediate economic distress, especially for the poor and vulnerable. Raising prices gradually while providing targeted social assistance can mitigate their impact. Phasing out subsidies following a preannounced schedule reduces policy uncertainty and allows consumers to smooth out adjustment costs over time. Scaling up existing social programs or implementing new ones can protect the poor from immediate price shocks. To offset price increases, efforts are also needed to improve efficiency on both the supply and demand side. Many countries have used energy-efficiency programs to ensure affordable energy for low-income households.

Putting a price on emissions would also prompt countries to move toward renewables and away from fossil fuel–powered electricity. Although new jobs and opportunities are created during the process, workers in communities reliant on the fossil fuel industry could experience massive social and economic disruptions. Retraining programs and strategies for pursuing greater economic diversification in the local economy are needed to ensure a just transition of the workforce.

OUTLINE OF THE REPORT

The report is organized as follows:

Chapter 1 presents an overview of power sector distortions in South Asia. It discusses the mechanisms and consequences of the three types of distortions in the upstream, core, and downstream segments of the power sector in all three countries.

Chapter 2 presents the methodological framework and theoretical foundation and illustrates how distortions are measured in practice. It also describes the main data sets used and discusses the limitations of the analysis.

Chapters 3–5 present country-specific analysis. They provide institutional background, illustrate analytical approaches, and present detailed estimation results for each country.

Chapter 6 addresses interactions across distortions and offers policy implications for power sector reform.

References

Badiani, Reena, and Katrina Jessoe. 2013. "The Impact of Electricity Subsidies on Groundwater Extraction and Agricultural Production." Working Paper, University of California, Davis.

BPDB (Bangladesh Power Development Board). 2015, 2016. *Annual Reports*. Dhaka.

Central Electricity Authority (CEA). 2000–2012. "Performance Review of Thermal Power Stations." New Delhi.

———. 2004. *Report of the Expert Committee of Fuels for Power Generation*. New Delhi.

———. 2015. "Annual Performance Review of Thermal Power Plants for 2014–2015." New Delhi.

———. 2018. *Load Generation Balance Report 2017-18*. New Delhi.

CERC (Central Electricity Regulatory Commission). 2015. *Annual Report* 2014-2015. New Delhi.

Chan, Ron, Edward Manderson, and Fan Zhang. 2017. "Energy Prices and International Trade: Incorporating Input-Output Linkages." Policy Research Working Paper WPS8076, World Bank, Washington, DC.

Coal India Limited (CIL). 2018. "Price Notification." New Delhi. https://www.coalindia.in /DesktopModules/DocumentList/documents/Price_Notification_dated_08.01.2018 _effective_from_0000_Hrs_of_09.01.2018_09012018.pdf.

Global Burden of Disease MAPs Working Group. 2018. *Burden of Disease Attributable to Major Air Pollution Source in India Special Report*. Boston.

Grainger, C. A. and Fan Zhang. 2017. "The Impact of Electricity Shortages on Micro- and Small-Enterprises: Evidence from India." Background paper for this report, World Bank, Washington, DC.

The Guardian. 2012. "Pakistan Power Cut Riots Spread as Politician's House Stormed." June 19. https://www.theguardian.com/world/2012/jun/19/pakistan-power-cut-riots.

Health Effects Institute. 2017. *State of Global Air 2017:* Boston.

Hydrocarbon Development Institute of Pakistan. 2001, 2007, 2014, 2016. *Pakistan Energy Yearbook.* Islamabad.

IEA (International Energy Agency). n.d. World Energy Statistics and Balances Database. Paris. https://www.iea.org/statistics/relateddatabases/worldenergystatisticsandbalances/.

———. 2013. Energy Subsidies Database. Paris.

IEA. 2017. *Energy Access Outlook 2017 From Poverty to Prosperity.* World Energy Outlook Special Report. Paris.

IFC (International Finance Corporation). 2015. *Pakistan Off-Grid Lighting Consumer Perception Study.* Washington, DC.

Indian Central Electricity Authority. 2017. *Growth of Electricity Sector in India from 1947 to 2017.* New Delhi.

Indian Power Finance Corporation. Various years. *Performance of State Power Utilities.* New Delhi.

India Planning Commission. Various years. *On the Working of State Power Utilities and Electricity Departments.* New Delhi.

IPCC (Intergovernmental Panel on Climate Change). 2006. *Guidelines for National Greenhouse Gas Inventories.* Geneva.

Jacobson, Arne, Tami Bond, Nicholas L. Lam, and Nathan Hultman. 2013. "Black Carbon and Kerosene Lighting: An Opportunity for Rapid Action on Climate Change and Clean Energy for Development." Policy Paper 2013-03, Brookings Institution, Washington, DC.

Joskow, Paul L. 2008. "Incentive Regulation and Its Application to Electricity Networks." *Review of Network Economics* 7 (4): 547–60.

Lam, Nicholas L., Yanju Chen, Cheryl Weyant, Chandra Venkataraman, Pankaj Sadavarte, Michael A. Johnson, Kirk R. Smith, Benjamin T. Brem, Joseph Arineitwe, Justin E. Ellis, and Tami C. Bond. 2012. "Household Light Makes Global Heat: High Black Carbon Emissions from Kerosene Wick Lamps." *Environmental Science & Technology* 46 (24): 13531–38.

Nikolakakis, T., Deb Chattopadhyay, Morgan Bazilian. 2017. "A Review of Renewable Investment and Power System Operational Issues in Bangladesh." *Renewable and Sustainable Energy Reviews* 68 (1): 650–58.

OECD (Organisation for Economic Co-operation and Development). 2015. *Inventory of Estimated Budgetary Support and Tax Expenditures for Fossil Fuels.* Paris. http://www.oecd.org/site/tadffss/.

Pakistan Ministry of Finance. 2017. *Pakistan Economic Survey 2016–17.* Islamabad

Pakistan National Transmission and Dispatch Company. 2016. Power System Statistics. Government of Pakistan, Islamabad.

Pargal, Sheoli, and Sudeshna Ghosh Banerjee. 2014. *More Power to India: The Challenge of Electricity Distribution.* Washington, DC: World Bank.

Petrobangla. 2015. *Annual Report 2014–2015.* Dhaka.

Press Information Bureau. 2017. *Year End Review 2017-MNRE.* Government of India, Delhi.

Rodell, M., I. Velicogna, and J. S. Famiglietti. 2009. "Satellite-Based Estimates of Groundwater Depletion in India." *Nature* 460: 999–1002.

Samad, Hussain, and Fan Zhang. 2016. "Benefits of Electrification and the Role of Reliability: Evidence from India." Policy Research Working Paper WPS7889, World Bank, Washington, DC.

———. 2017. "Heterogeneous Effects of Rural Electrification: Evidence from Bangladesh." Policy Research Working Paper WPS8102, World Bank, Washington, DC.

———. 2018. "Electrification and Household Welfare: Evidence from Pakistan." Policy Research Working Paper WPS8582. World Bank, Washington, DC.

Schwab, Klaus. 2018 *The Global Competitiveness Report 2017–2018.* Geneva: World Economic Forum.

Sekhri, S. 2013. "Missing Water: Agricultural Stress and Adaptation Strategies in Response to Groundwater Depletion among Farmers in India." Working Paper. International Growth Centre, London School of Economics and Political Science, London.

Shindell, Drew, Johan C. I. Kuylenstierna, Elisabetta Vignati, Rita van Dingenen, Markus Amann, Zbigniew Klimont, Susan C. Anenberg, Nicholas Muller, Greet Janssens-Maenhout, Frank Raes, Joel Schwartz, Greg Faluvegi, Luca Pozzoli, Kaarle Kupiainen, Lena Höglund-Isaksson, Lisa Emberson, David Streets, V. Ramanathan, Kevin Hicks, N. T. Kim Oanh, George Milly, Martin Williams, Volodymyr Demkine, and David Fowler. 2012. "Simultaneously Mitigating Near-Term Climate Change and Improving Human Health and Food Security." *Science* 335 (6065): 183–89.

Tedsen, E. 2013. *Black Carbon Emissions from Kerosene Lamps: Potential for New CCAC Initiative.* Ecological Institute, Berlin.

USAID (U.S. Agency for International Development). 2011. *Evaluation of Economic Value of Natural Gas in Various Sectors.* Washington, DC: USAID.

———. 2013. *The Causes and Impacts of Power Sector Circular Debt in Pakistan.* Study commissioned by the Planning Commission of Pakistan. Washington, DC.

Veettil, Prakashan Chellattan, Stijn Speelman, Aymen Frija, Jeroen Buysse, Koen Mondelaers, and Guido van Huylenbroeck. 2011. "Price Sensitivity of Farmer Preferences for Irrigation Water–Pricing Method: Evidence from a Choice Model Analysis in Krishna River Basin, India." *Journal of Water Resources Planning and Management* 137 (2): 205–14.

WHO (World Health Organization). 2015. *Reducing Global Health Risks Through Mitigation of Short-Lived Climate Pollutants: Scoping Report for Policymakers.* Geneva: WHO.

World Bank. n.d. Global Economic Monitor Commodities Database. Washington, DC.

———. Various years. *Enterprise Surveys.* Washington, DC.

———. 2015a. *Pakistan: Second Power Sector Reform Development Policy Credit Program Project Document.* Washington, DC.

———. 2015b. "A Review of Renewable Investment and Power System Operational Issues in Bangladesh." Also published in 2017 in *Renewable and Sustainable Energy Reviews* 650–58.

———. 2018. World Development Indicators Database. Washington, DC. http://databank.world bank.org/data/reports.aspx?source=world-development-indicators.

CHAPTER 1

What Are the Distortions?

Over the past decade, countries in South Asia have made enormous progress in expanding access and supply of electricity. However, in absolute terms, South Asia still faces massive electricity shortages. About 255 million people in the region lack access to electricity, or more than a quarter of all those living off the grid globally. For firms and households that are connected to the grid, electricity supply is often erratic, with frequent and long power outages. Outages often occur because of technical failures. In South Asia, they also reflect the efforts of utilities to cope with persistent power shortages through scheduled blackouts, known as load shedding.

According to the most recent World Bank Enterprise Surveys conducted during 2011–15, business managers in South Asia reported that power cuts occur almost daily, lasting on average 5.3 hours. By comparison, business managers in East Asia reported one outage every nine days, and managers in Sub-Saharan Africa reported one every four days. The 2018 *Global Competitiveness Report* ranks Bangladesh, India, and Pakistan 101th, 80th, and 115th among 137 economies in the reliability of electricity supply (Schwab 2018). As a result of low access rates and the low quality of supply, South Asia has the world's second-lowest per capita electricity consumption. At 707 kilowatt-hours a year, it is less than a quarter of the world average, according to the World Bank's World Development Indicators. Meanwhile, increasing dependence on fossil fuel–based power generation has contributed to the worsening air pollution in South Asia; it is now a region with some of the most polluted cities in the world.

Conventional wisdom says that inadequate investment in the power infrastructure is the main cause of power shortages in South Asia. But a closer look at the data reveals a different picture. Over the decade ending in 2016, Bangladesh and India more than doubled their power generation capacity, with average annual growth in capacity outstripping annual growth in their gross domestic product (GDP). But in Bangladesh, less than 80 percent of available capacity was operational most of the time (BPDB 2015, 2016). In India, power shortages reached 5 percent of estimated demand in 2014, while up to 15 percent of coal power plants were left idle (CERC 2015). In fiscal 2018, when total installed capacity was more than twice the amount of peak demand, peak demand shortage still registered 2.1 percent in India (CEA 2018). Even in Pakistan, where capacity growth lags GDP growth, only 80 percent of available capacity was operational in fiscal 2014 (World Bank 2015a). Losses in transmission and distribution add to the shortages: India and Pakistan lose about a quarter of electricity in the network, well above the 10 percent international norm.

South Asia thus faces an efficiency gap. Inefficiencies in every link of the electricity supply chain have resulted in upstream fuel shortages, poorly performing state utilities, and wasteful consumption downstream, all contributing significantly to the supply deficit. Although there are multiple inefficiencies, most are attributable to three types of distortions: institutional distortions caused by state ownership and weak governance; regulatory distortions resulting from price regulation, subsidies, and cross-subsidies; and social distortions relating to the negative externalities (such as environmental and health costs) of energy production and consumption.

Addressing each type in turn, this chapter describes the mechanisms and consequences of distortions along the electricity supply chain in the three largest economies in South Asia: Bangladesh, India, and Pakistan. These countries have a combined population of 1.6 billion, including almost 300 million people living in extreme poverty (that is, subsisting on less than $1.90 a day) and 245 million lacking access to electricity. The three countries account for 98 percent of South Asia's electricity supply. The power sectors of the three countries together emitted 1.15 billion tons of carbon dioxide in 2015, almost as much as the power sectors of all of the Organisation for Economic Co-operation and Development (OECD) member countries in Europe (IEA 2017a). The distortions described in this chapter are by no means exhaustive. But they do illustrate the most common and important sources of inefficiencies in the power sectors of these countries.

Institutional Distortions

Institutional distortions in South Asia's power sector arise from the interplay of government ownership and lack of competition in energy production and supply. Despite recent reforms, the fuel and electricity sectors in South Asia are still heavily regulated and mostly managed by public institutions.

Government-owned entities typically lack a strong profit incentive to improve efficiency. More important, they are susceptible to political interference. When a government pursues political goals that are inconsistent with maximizing social welfare, both productive and allocative efficiencies are harmed (Shleifer and Vishny 1994). For example, a government may channel benefits to political supporters by providing excessive employment and high wages at state-controlled companies or unfairly favoring some plants over others in the allocation of resources.

The inefficiency associated with public ownership is compounded by lack of market competition. Economists have long maintained that a competitive market generates an efficient allocation of resources. Markets can also promote efficiency by revealing information about managerial effort (Laffont and Tirole 1993). In a competitive power market, for example, inefficient power plants will be dispatched the least and make the lowest profits. Whether monitored by the regulator or not, their managers will be forced to improve efficiency. In this way, the market functions as an effective monitoring mechanism. Where the power market is not competitive, customers tend to bear the burden of inefficient behavior by managers, with higher operating costs generally reflected in higher rates.

UPSTREAM: UNPRODUCTIVE MINES, LEAKING PIPELINES, AND PRIVILEGED ACCESS

Coal and natural gas play key roles in the fuel mix for power generation in South Asia, and so their production is a critical link in the power supply chain. In 2015 coal accounted for 75 percent of electricity production in India, and gas accounted for more than 80 percent in Bangladesh and 27 percent in Pakistan. India holds the world's third-largest coal reserves, yet supply has still fallen short of demand in recent years. And, despite having historically large reserves of gas, both Bangladesh and Pakistan face dwindling reserves and significant shortages of domestic gas for power generation today. Institutional structures that emphasize incontestable public monopolies have led to an inefficient fuel supply and exacerbated upstream fuel shortages.

In India, coal mining is dominated by two government-owned enterprises, one of which controls more than 80 percent of production and distribution. Lacking competitive pressure from private commercial mining, both companies have been slow to adopt technologies aimed at increasing safety and efficiency. One telling sign is the extremely low automation level in underground mining. Less than 12 percent of the 252 underground coal mines across India were mechanized as of fiscal 2016 (CIL 2016). Labor productivity as measured by output per labor shift has remained nearly stagnant for decades, and labor use is astonishingly intensive even by Indian standards: underground mining contributed less than 10 percent of output but employed half the workforce in the coal sector (Figure 1.1).

In Bangladesh and Pakistan, the gas sector has long been open to private investors, but the government still wields substantial influence over its operation and pricing. In Pakistan, for example, gas exploration and production are open to competition from the private sector, but two mostly government-owned companies control all gas transmission and distribution. The government intervened in the operation of both by encouraging rapid extension of gas connections to residential consumers. As a result, the length of the transmission network in the country has more than tripled since 1996, making it the most extensive inland gas supply system in the world, long enough to circle the earth more than three times. Meanwhile the two companies, each a monopoly in its

FIGURE 1.1 Inadequate technology contributes to declining output from underground coal mines in India

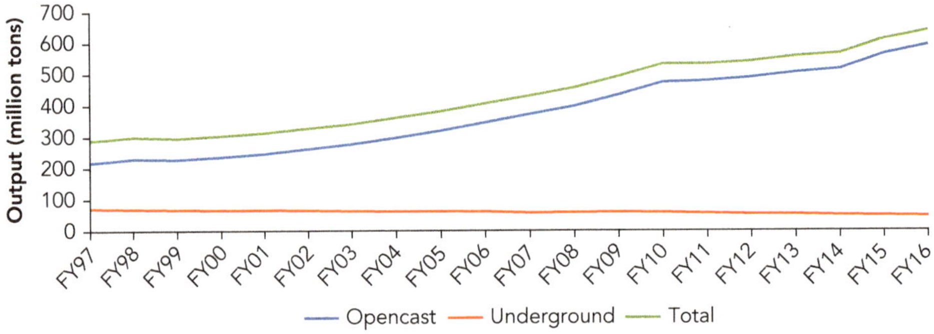

Source: Ministry of Coal, *Provisional Coal Statistics*, accessed through Indiastat.
Note: FY = fiscal year.

own market, are allowed to pass their operating costs on to consumers, thereby undermining financial incentives to reduce gas losses through pipeline inspection and repair. The rapid network expansion, combined with the neglect of both pipeline maintenance and theft prevention, has led to increasingly high volumes of gas losses. In 2015 a staggering 14 percent of gas was lost in the network. By comparison, the standard rate of gas loss in OECD countries is a mere 2 percent (World Bank 2015b).

Lack of market mechanisms has implications well beyond mismanagement at the firm level. It also undermines allocative efficiency, especially when resources are allocated on the basis of political rather than economic considerations. In the wake of acute gas shortages, the governments of both Bangladesh and Pakistan began to ration gas to different sectors. Under normal conditions, the market can effectively solve the allocation problem: consumers able to extract more value from gas are willing to pay a higher price, leading to optimal allocation of the resource. But, in Bangladesh and Pakistan, the allocation of gas is determined by administrative orders and often favors less productive uses.

In Bangladesh, for example, more efficient power plants are more often affected by gas shortages, even after taking into account differences in the location, size, and age of plants (Figure 1.2). In Pakistan, power generation ranked third in priority for gas allocation before 2012—even though that sector is estimated to produce the greatest medium-term economic benefit of its use (USAID 2011). And, despite a policy change in 2012 that raised its priority ranking, the power sector was allocated only 33 percent of gas in fiscal 2016.

India faces similar inefficiency in resource allocation. Not long ago, a coal allocation scam made headlines in the country. In the aftermath of the scandal, India passed the Coal Mines (Special Provisions) Act, 2015, which stipulates that all coal blocks (coal leases) must be allocated by auction. But government-owned companies are exempt from competitive bidding; they are guaranteed an allotment under a separate window.

FIGURE 1.2 More efficient power plants are more likely to be affected by gas
shortages than less efficient plants in Bangladesh

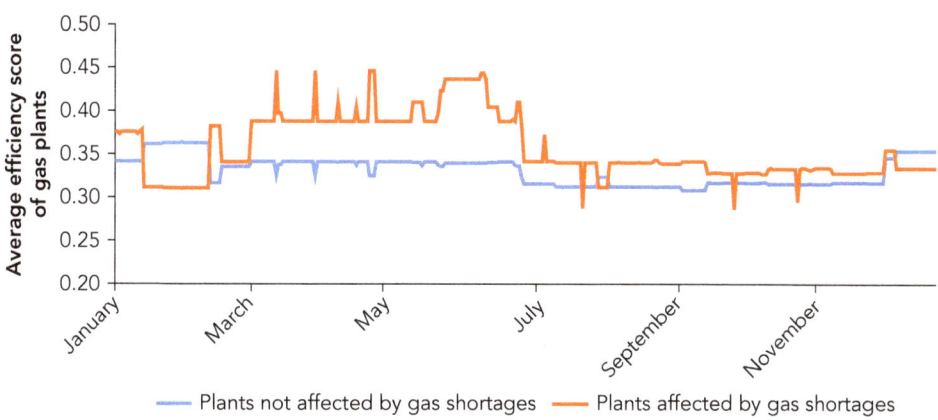

Source: Based on daily general reports, Bangladesh Power Development Board, January 1–December 31, 2014.
Note: Efficiency score is the ratio of electricity output to gas input in calorific value. Average efficiency is weighted
by capacity.

Meanwhile, private power plants, which are generally more efficient than state
government–owned power plants, often struggle to maintain a steady supply of coal.
Resource allocation that fails to reward efficiency could compromise the overall fuel
productivity of India's power sector.

CORE: INEFFICIENT GENERATION, HIGH LOSSES, AND FAVORITISM IN DISPATCH

Historically, the core sector of electricity supply—generation, transmission, and
distribution—has been considered a natural monopoly and operated by a single utility
to exploit economies of scale. But advances in technology in recent decades have led to
big changes in the institutional structure of electricity sectors around the world. Thanks
to the development of efficient generation technologies, the generation segment is no
longer considered a natural monopoly. As a result, countries have undertaken reforms
aimed at fostering competition in generation and changing the way transmission and
distribution are regulated to ensure a level playing field for all participants.

With the aim of attracting private investment, Bangladesh, India, and Pakistan have all
followed the global trend of power sector reform since the 1990s. Bangladesh and Pakistan
unbundled their national power companies into separate generation, transmission, and
distribution entities. In India, the central government mandated the unbundling of the
state-level vertically integrated utilities, the state electricity boards. Implementation has
been uneven, however: 18 states and the National Capital Territory of Delhi completed
the unbundling to varying degrees, but 11 states are still operating their utilities as a
single entity.

The unbundling was accompanied by policies to spur private participation. Private generators now account for 43 percent of generation capacity in Bangladesh, 44 percent in India, and 45 percent in Pakistan. In distribution, by contrast, privatization remains limited. In Bangladesh, the distribution sector was restructured into several companies, but all of them are still owned and operated by the government. In India, the National Capital Territory of Delhi and the state of Odisha fully privatized the distribution sector in the early 2000s, and seven other states set up private distribution companies. Odisha's distribution utilities were returned to government control in 2015 following poor performance over a long period. Privately operated distribution currently accounts for about 6 percent of electricity sales in the country. Pakistan privatized only one distribution company, the Karachi Electric Supply Company, which accounts for about 14 percent of electricity distribution. All three countries have kept transmission and dispatch mostly under government control.

Power sector reform has led to greater private investment, but it has not fully enabled competition and has yet to establish an effective incentive structure. In India, only 4 percent of electricity was exchanged through the competitive wholesale market in fiscal 2016 (CERC 2017). And, in all three countries, most generators are paid under long-term power purchase agreements following rate-of-return regulation.

Because generators are shielded from market competition and costs are passed on to consumers, incentives to improve efficiency remain weak, especially for publicly owned power plants. Controlling for differences in plants' physical and technical characteristics (such as age, capacity, technology, location, and dispatch), analysis in this report reveals a large efficiency gap between public and private power plants. The difference is both statistically strong and economically nontrivial. Compared with an independent power producer, a state government–owned coal power plant in India would use 16 percent more fuel on average to produce the same amount of electricity. In Bangladesh, a government-owned gas plant would use 29 percent more fuel and in Pakistan 20 percent more.

Analysis based on total factor productivity shows a similar efficiency gap (Figure 1.3). Because the analysis controls for power plants' exogenous physical and operational characteristics and their observable inputs, the remaining differences in operating efficiency between public and private power plants are likely caused by differences in the quality of management practices, although some of this difference could also be explained by the type of power purchase agreements signed by private plants, which allows them to be dispatched at optimal load factors.

Compounding the effects of inefficient generation, around a fifth of the electricity produced in India and Pakistan is lost in the transmission from supply sources to distribution points and in the distribution to consumers. This share is substantial, especially for countries already struggling to bridge the gap between supply and demand. Figure 1.4 puts these losses in context. For both countries, the rate of transmission and distribution losses is substantially higher than the world average, which has remained below 10 percent for the past two decades. It also exceeds the average for

FIGURE 1.3 Public power plants are substantially less efficient than private ones: Bangladesh and Pakistan as examples

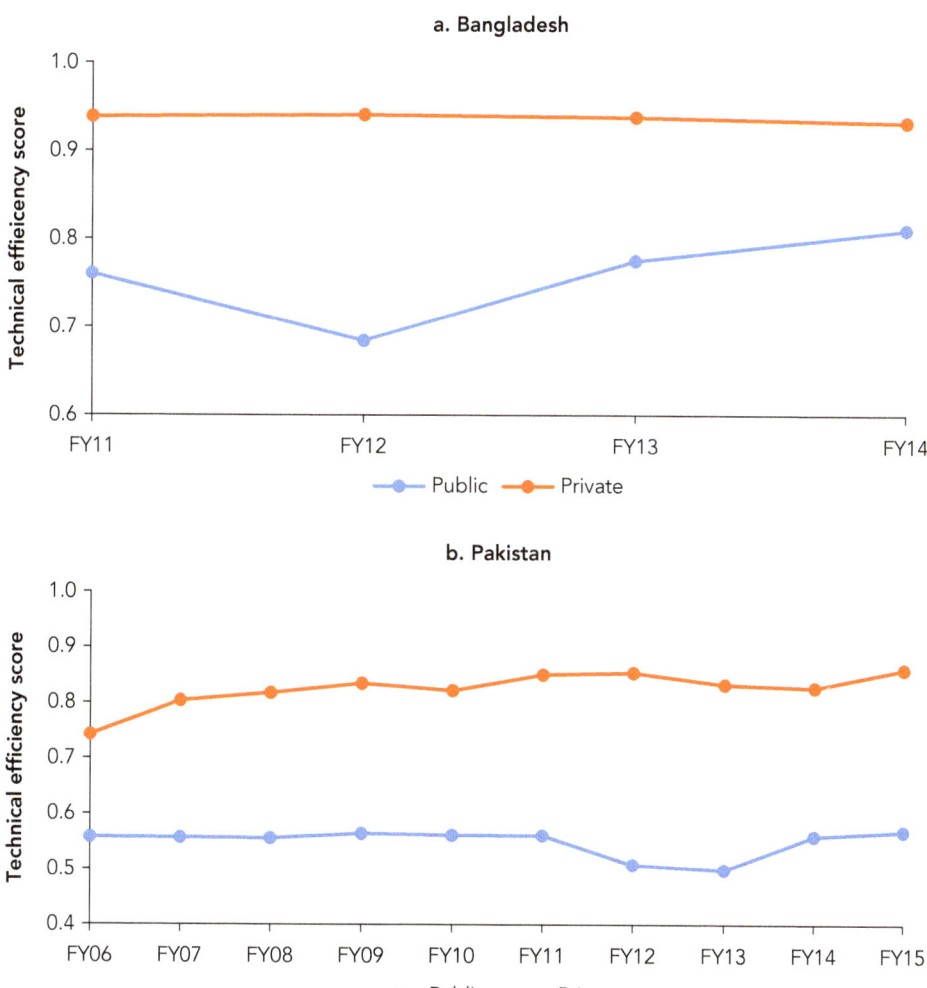

Source: Estimation based on plant-level data from Bangladesh Power Development Board (BPDB) annual reports (2011–14) and the Pakistan National Electric Power Regulatory Authority's (NEPRA) State of Industry Report (2006–15). *Note:* Technical efficiency score measures the ratio of actual output to maximum feasible output. Private plants refer to independent power producers but not rental power plants in Bangladesh. FY = fiscal year.

lower-middle-income countries. These losses are caused by a combination of electricity theft, poor infrastructure, faulty metering, outdated equipment, and other factors.

In India, high losses in distribution, compounded by poor recovery of overdue electricity bills, has led to alarming levels of debt and repeated government bailouts. India's central government has launched three rescue operations to bail out loss-making distribution companies. In 2001–02 it assumed 350 billion rupees (Rs) in debt and waived 50 percent of the outstanding interest. In 2012 it issued a bailout package worth

FIGURE 1.4 Transmission and distribution losses substantially exceed the world average in India and Pakistan

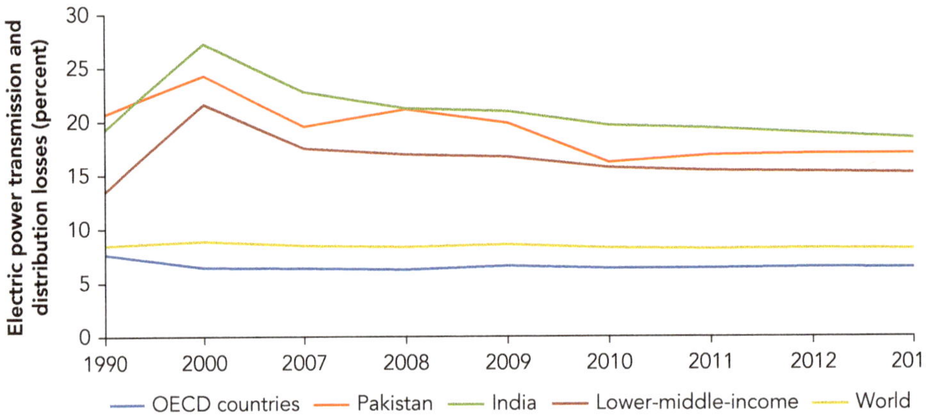

Source: World Bank (2018).
Note: OECD = Organisation for Economic Co-operation and Development.

Rs 1.9 trillion, or more than four times the size of the first. By March 2015, distribution companies were again in trouble, with accumulated losses of roughly Rs 3.8 trillion and outstanding debt of almost Rs 4.3 trillion, at interest rates as high as 15 percent. The latest bailout package, Ujwal DISCOM Assurance Yojana, was announced at the end of 2015. State governments took over 75 percent of the outstanding liabilities of distribution companies.

Pakistan's government has periodically paid down the "circular debt" accumulated from the combined high losses in the transmission and distribution sectors—a debt that had reached a staggering $9 billion by the end of fiscal 2012 (USAID 2013). Repeated large-scale government rescues create a classic situation of soft budget constraint (Kornai 1979), weakening incentives to improve efficiency and further softening the budget constraint.

Institutional distortions also show up in the inefficient dispatch of electricity. An efficient dispatch system follows a merit order: at any given time, the system operator meets demand by taking electricity first from the least expensive power plants and then from the next less expensive ones, until all demand is met. In Bangladesh, however, oil-based rental power plants are often dispatched before gas-based plants, despite an average cost for the former that is 5–12 times higher and an emission intensity that is from 30 to 600 percent greater, according to a World Bank study (World Bank 2015c). Technical constraints such as transmission congestion could contribute to out-of-merit dispatch. But dispatch can also be vulnerable to political interference where there is heavy government involvement in the core sector. The same study finds that Bangladesh could reduce its production costs by 63 percent through market measures to enhance the efficiency of dispatch (World Bank 2015c).

DOWNSTREAM: LOWER LIVING STANDARDS AND SLOWER BUSINESS GROWTH

Almost 255 million people in South Asia lacked access to grid electricity in 2016, including 205 million in India, 39 million in Bangladesh, and 1.6 million in Pakistan. Most of them lived in rural areas. Together, they represent more than a quarter of the global population living off the grid (Table 1.1).

Lack of access to electricity diminishes the quality of life. It prevents households from using electrical appliances that improve comfort and convenience, such as fans and refrigerators, and electronic devices that provide access to information, such as television. Without electricity, It is more difficult for children to study in the evening and for shops to remain open at night.

What is the opportunity cost of lack of access to electricity? To quantify this cost, the analysis delves into household survey data, comparing over an extended study period changes in income, education, and employment outcomes for households following grid connection and households remaining off the grid. The results reveal that in all three countries gaining access to the grid is associated with substantial improvements in income, employment, health outcomes, and time spent studying. Nonfarm income accounts for most of the income growth, suggesting that electricity may provide opportunities for more diversified economic activities in rural areas.

Gaining access to electricity can especially benefit women and girls, who traditionally are responsible for household chores. Indeed, the analysis shows that access reduces the time they spend collecting biomass fuel by 44 percent in India. In addition, gaining access to electricity increases girls' study time in Bangladesh and India and women's labor force participation in all three countries. By contrast, the analysis finds no statistically significant positive effects on men's employment or labor force participation in the three countries.

Electrification also boosts gender equality through both greater economic empowerment and better access to information for women. In households with an electricity connection, women have more decision-making power over education (both their own and their children's), health care, and purchases, as well as greater access to assets and financing (Samad and Zhang 2018).

TABLE 1.1 South Asia has low rates of access to electricity, especially in rural areas

Country	Total (percent of population)	Rural (percent of population)
Bangladesh	75.9	68.9
India	86.1	81.0
Pakistan	99.1	98.8

Source: World Development Indicators database. Indian data are from the Indian rural electrification program's (Saubhagya) dashboard, updated as of October 2017.
Note: Data are for 2016 except for India.

Another important dimension of the welfare effects of electrification is the reliability of electricity supply. Power shortages constrain households' use of electricity. And even occasional outages may discourage investment in electrical appliances or a sustained change in study or work patterns (Banerjee and others 2015). In all three countries, power outages have a negative effect on income, study time, employment, and almost all other development outcomes considered (Figure 1.5).

An unreliable power supply also has enormous consequences for firms, reducing their productivity and thus their output and profits. According to the most recent World Bank Enterprise Surveys, conducted between 2011 and 2015, more than a fifth of firms in South Asia identify unreliable electricity supply as the biggest barrier to business growth. Indeed, it tops a list of 15 constraints that include access to finance, corruption, taxes, and human capital.

Power outages affect industrial productivity and long-term growth in several ways. First, outages may force firms to invest in expensive generators, thereby diverting capital from more productive uses. When firms lack an alternative source of electricity,

FIGURE 1.5 **Access to electricity is associated with higher income and better social outcomes in Bangladesh, India, and Pakistan—and the results are much stronger if the electricity is reliable**

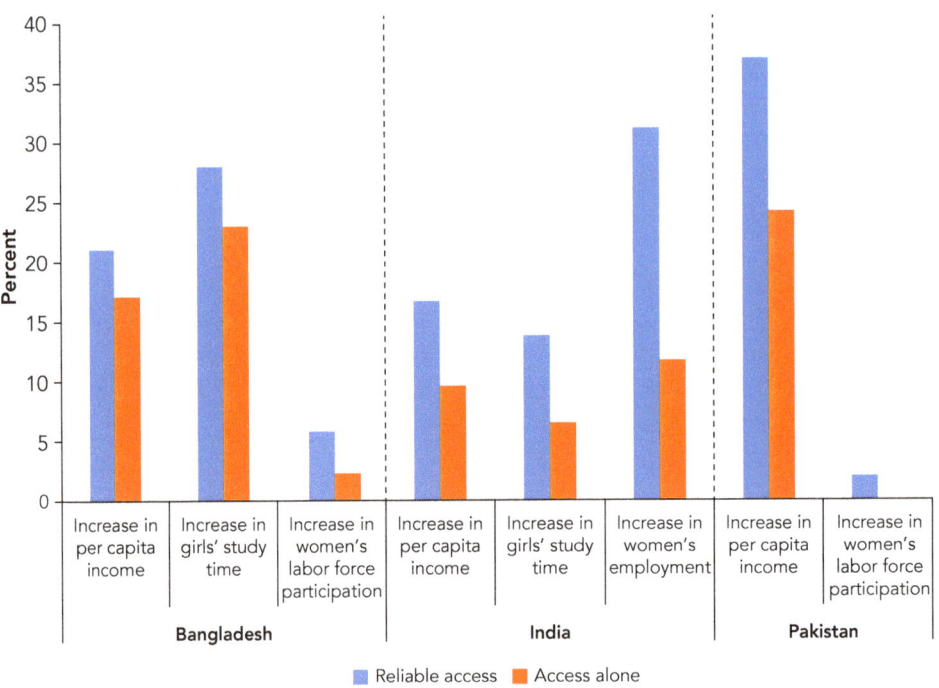

Source: Estimation based on household surveys in Bangladesh, India, and Pakistan. See also Samad and Zhang (2016, 2017, 2018).
Note: The effects of electrification on girls' study time and the effects of power outages on women's labor force participation in Pakistan are not estimated because data are not available.

particularly for unanticipated outages, they must shut down and send workers home, which reduces the productivity of labor. Second, firms facing power shortages may substitute away from electricity-intensive technology. Because production processes that are less electricity-intensive also tend to be less technologically advanced, switching to such processes undermines long-term productivity growth for firms (Abeberese 2017). Third, firms may choose to substitute for electricity by purchasing intermediate inputs, significantly increasing their unit production costs (Fisher-Vandern, Mansur, and Wang 2015).

Electricity shortages result in substantial losses for businesses. Analysis in this report shows that in Bangladesh a 10 percent power shortage is associated on average with a 3.1 percent reduction in total factor productivity for manufacturing firms. In India, a 7.3 percent power shortage reduces the revenue and producer surplus of large firms by 5.6–7.7 percent on average (Allcott, Collard-Wexler, and O'Connell 2016). In Pakistan, a one-hour increase in the average daily duration of scheduled power outages leads to a reduction in a firm's value added and revenue of roughly 1.3 percent (Grainger and Zhang 2017a).

The average effect of power shortages is large, and the effect on micro- and small enterprises is even larger. Analysis using data from the Fourth All-India Census of Micro, Small, and Medium Enterprises finds that, for smaller firms with a median size of two employees, the effects are almost two orders of magnitude greater than those for larger firms with an average of 167 employees (Grainger and Zhang 2017b). The most likely reason is that smaller firms are much less likely to purchase generators to cope with power shortages. Indeed, according to India's Annual Survey of Industries, self-generation increases sharply with plant size: about 75 percent of plants with more than 500 employees have self-generation capacity, whereas only 10–20 percent of plants with fewer than 10 employees have such a capacity (Allcott, Collard-Wexler, and O'Connell 2016).

Micro-, small-, and medium-size enterprises play an important role in most economies, particularly in developing ones (Li and Rama 2015). In India, informal firms, which are often the smallest, account for about 75 percent of manufacturing employment (Hsieh and Klenow 2014); and micro-, small-, and medium-size enterprises are estimated to contribute as much as 45 percent of GDP (World Bank 2015c). The particularly large impact of power shortages on smaller enterprises could therefore lead to substantial losses in jobs and growth. It could also result in the misallocation of resources between small and large firms—and thus in lower total factor productivity for the industrial sector as a whole (Allcott, Collard-Wexler, and O'Connell 2016).

Regulatory Distortions

Regulatory distortions in the South Asia power sector arise from the mispricing of different energy products. Mispricing includes subsidies for fuel and electricity, cross-subsidies among consumer groups, and pricing arbitrage between fuel and electricity, which encourages inefficient captive power generation. This report defines subsidies for coal as the difference between the regulated price for power production and the market-clearing price; subsidies for gas as the difference between prices for domestic gas and the (higher) international price; and subsidies for electricity as the difference between the average end-user price and the average supply cost.

By these definitions, subsidies do not always incur large direct budgetary costs. In Bangladesh and Pakistan, for example, prices for domestic gas are close to the cost of production. Yet, by selling gas at prices below the international market price, the government forgoes substantial revenues that could be invested in economic growth in the longer term. Subsidies also encourage wasteful consumption, result in an economically inefficient allocation of resources, and deter investment in upstream exploration and production. Evaluated by their opportunity costs, fuel subsidies are much larger than electricity subsidies in South Asia.

UPSTREAM: UNDERPRICED COAL AND GAS

Coal and gas are priced substantially below their opportunity cost in South Asia, even without factoring in their external costs to the environment. In India, although coal is not directly subsidized, the price of coal for the so-called regulated sectors—power, fertilizer, and defense—is lower than the price charged for all other sectors—as a way to keep electricity prices low. In 2018 the price of medium-quality coal (grade G9) was 17 percent lower for power generation than for consumers in unregulated sectors (CIL 2018). As a price discovery mechanism, about 10 percent of coal is sold on the spot market rather than through long-term purchase agreements. The spot market allows consumers in all sectors to purchase coal through an electronic auction by making bids at or above the reserve price. In any given year, the spot market price is substantially higher than the discounted rate charged to power producers (Figure 1.6). In fiscal 2016, for example, power utilities would have had to pay roughly 37 percent more for G9 coal if they had bought their coal through the electronic auction.

In Bangladesh and Pakistan, domestic natural gas has been consistently priced below the import parity price—the landed price of liquefied natural gas (LNG) at the nearest international hub (Figure 1.7). Even with the recent drop in global gas prices, the international LNG price in fiscal 2016 was still almost 11 times the domestic gas price for power generation in Bangladesh and 2.7 times the corresponding domestic gas price in Pakistan. Gas underpricing can also be evaluated against the cost of the cheapest replacement fuel. For power generation, this is the price of furnace oil, which in fiscal 2017 was 19 times the price of gas for power generation in Bangladesh and 2.1 times the price of domestic gas in Pakistan.

The low price of domestic fuel contributes to wasteful consumption and missed opportunities for diversifying the fuel mix for power generation. In Bangladesh, for example, low gas prices have led to a bias toward gas-intensive power production. The efficiency of gas-based generation in Bangladesh is ranked the 15th lowest in the world. Meanwhile, gas has emerged as the only major fuel for electricity generation,

FIGURE 1.6 The price of coal for power generation is much lower than the spot market price in India

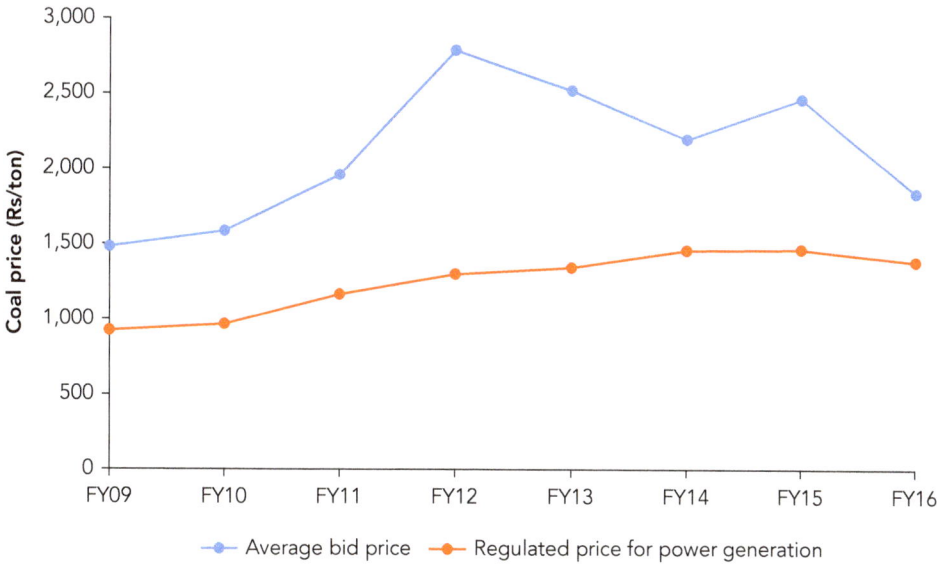

Source: Ministry of Coal, India (2017).
Note: FY = fiscal year.

FIGURE 1.7 The price of domestic natural gas is much lower than the international price in Bangladesh and Pakistan

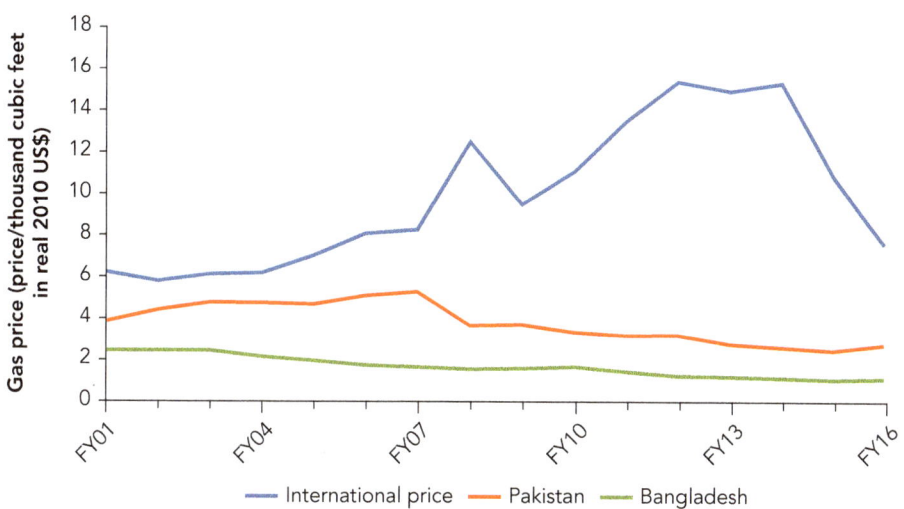

Source: Petrobangla (2016); Hydrocarbon Development Institute of Pakistan (2001, 2007, 2014, 2016); Ministry of Finance, Pakistan (2017); World Bank Global Economic Monitor Commodities database.
Note: FY = fiscal year.

with its share in the fuel mix rising from 40 percent in 1971 to a peak of 93 percent in 2010 (before falling back to 81 percent after the onset of the gas shortage). The government has developed an ambitious fuel diversification plan that envisages reducing the share of gas to 25 percent of the total energy supply by 2030. Yet state utilities continue to build inefficient gas power plants. These plants consume on average 11–12 cubic feet of gas per kilowatt, or almost twice the 6–7 cubic feet for a typical combined-cycle power plant.

On the supply side, low profitability discourages private investors from carrying out exploration activities. Several large gas development projects have been abandoned in Bangladesh because of the government's unwillingness to raise tariffs to levels that would enable cost recovery with reasonable returns (Rahman 2015).

Another form of regulatory distortion is cross-subsidies among consumer groups. In the upstream coal sector, India's rail system cross-subsidizes passenger service with coal freight, making India's freight rates (adjusted for purchasing power parity) among the highest in the world (Government of India 2015). Because coal must be hauled from mines located mainly in eastern states to power plants scattered across the country, the rail system plays a critical part in coal supply. Railways delivered about 274 million tons of coal to power plants in fiscal 2014, over an average distance of 542 kilometers (CEA 2015). The cross-subsidization of passenger rail service by freight rail service in India not only increases transport costs for coal and the end price of electricity but also undermines efficiency and investment in freight rail, creating major bottlenecks in coal supply. In 2014 railway capacity constraints, including line congestion and a shortage of railcars, left 50 million tons of coal to be delivered by railway stranded at mines (EIA 2015).

CORE: UNDERPRICED ELECTRICITY AND INEFFICIENT TRANSMISSION PRICING

The retail price of electricity for households and farmers in Bangladesh, India, and Pakistan is substantially lower than the cost of supply. The gap is financed through cross-subsidization and direct government subsidies. In Bangladesh, the budgetary support for electricity in fiscal 2016 amounted to $780 million. In India, the biggest chunk of subsidies goes to the agricultural sector in the form of heavily subsidized electricity from the grid for running irrigation pumps. The weighted-average tariff across all categories of consumers is 22 percent lower than the average unit supply cost, leading to a budget allocation for electricity subsidies of $8.8 billion in fiscal 2016 (Indian Planning Commission 2014; Power Finance Corporation 2017). In Pakistan, recent electricity tariff reforms lowered subsidy spending from 1.2 percent of GDP in fiscal 2012 to 0.8 percent in fiscal 2015. But direct subsidies for electricity in fiscal 2015 still cost the national exchequer $2.15 billion.

Low electricity prices have long been viewed as a way to expand the access of poor households to electricity. But they lead to unintended consequences. One problem is that governments do not always pay subsidies in time. Along with irregularities in billing and the rampant theft of electricity, this problem has contributed to the financial difficulties of distribution utilities, resulting in underinvestment and creating perverse incentives for utilities to underserve loss-making customers. Using data from the electricity sector in Colombia, McRae (2015) finds that subsidies deter investment in modernizing infrastructure and trap households and utilities in a nonpayment, low-quality equilibrium.

This finding is consistent with analysis for this report using nighttime satellite images from India in 2013. These images show that areas adjacent to villages electrified between 2005 and 2012 experienced worse power outages in 2013, possibly because, as more low-paying consumers joined the grid, distribution utilities were either unable or unwilling to invest in maintaining and upgrading infrastructure to expand the power supply (Box 4.3).

Underpricing and nonpayment are also the main contributors to circular debt in the power sector. When distributors in Pakistan became insolvent, they could not pay the generators. The generators in turn could not pay fuel suppliers, leading to fuel supply shortages. In fiscal 2014, circular debt led to the idling of up to 5 gigawatts of capacity, or almost 22 percent of installed capacity in the country (World Bank 2015a).

Another pricing issue is related to transmission, where mispricing contributes to electricity shortages. Economic theory has established that the first-best approach to pricing electricity is to ensure that at any point (or node) on a network the price equals the marginal cost of providing electricity at that node. This cost includes both generation and delivery, taking into account transmission constraints and losses. This type of pricing sends a proper signal for investment: areas experiencing more frequent transmission constraints face higher transmission charges, which should attract more investment in transmission cables or justify local generation. A simpler way of charging is to impose transmission rates that charge for the maximum capacity paid by the user instead of the generator.

Chile, New Zealand, and some power pools in the United States have adopted this nodal pricing approach. Other countries use zonal pricing, a simplified form of nodal pricing that groups nodes into zones and applies different prices to different zones based on the principle of marginal cost pricing. India introduced point of connection–based transmission pricing, making transmission charges distance and direction sensitive. Most other South Asian countries, however, still see transmission as an overhead cost and charge simple "wheeling rates" for the use of transmission lines, with payments determined by the volume of electricity transferred and the length of the contracted routes. This pricing approach does not take into account the scarcity of transmission capacity, even in areas frequently experiencing congestion. For example, in Pakistan transmission constraints accounted for 29 percent of the electricity shortfall in fiscal 2015. Meanwhile, the state-run dispatch company could not prioritize investments for improving grids where most congestion occurred (NEPRA 2016).

DOWNSTREAM: CROSS-SUBSIDIES PENALIZING COMPETITIVENESS

Low electricity prices for households and farmers in South Asia are financed in part through cross-subsidies from industrial and commercial consumers. In Bangladesh in September 2015, small industrial firms were paying twice as much for electricity as agricultural users. In India, the industrial tariff in fiscal 2015 was almost three and a half times the price paid by farmers (Figure 1.8). In Pakistan, the average industrial tariff in 2013 was 37 percent higher than the residential tariff. These cross-subsidy rates are high because industrial customers are typically less costly to serve than small-scale residential and agricultural consumers. In developed countries, industrial users normally pay lower electricity prices than other types of users.

Because electricity is an essential input to production, higher industrial tariffs can mean higher prices for almost all goods and services in the domestic market. In developing and developed countries alike, low-income households are disproportionately affected because energy-intensive goods such as food typically account for a larger share of their budget (Grainger and Kolstad 2010; Grainger, Zhang, and Shreiber 2015; Kerkhof and others 2008).

By increasing the cost of production, high industrial tariffs can also undermine export competitiveness. The effect can be especially strong in countries such as India that are less integrated in the global supply chain and therefore rely more on the domestic market for intermediate inputs. In these countries, higher electricity tariffs can have a multiplier effect as the impact accumulates along the supply chain, raising the cost of both intermediate and final production.

Like frequent power outages, high electricity tariffs induce firms to produce goods that are less electricity-intensive. Doing so has negative implications for long-run productivity growth because industries with initial low electricity intensity tend to later experience lower productivity growth (Abeberese 2017).

FIGURE 1.8 Industrial users cross-subsidize agricultural and residential consumers of electricity in Bangladesh, India, and Pakistan

Source: BPDB (2015); Indian Planning Commission (2015); National Transmission and Distribution Center (2014). Consumer Price Index (CPI) data: Ministry of Finance, Bangladesh (2015); Ministry of Finance, India (2017).
Note: FY = fiscal year.

In addition, when industrial electricity tariffs are high relative to fuel prices, firms may engage in inefficient captive generation. In Bangladesh, for example, the cost of electricity from gas-based self-generation, even with no waste heat recovery, is lower than the electricity tariff for industrial users (ADB 2014). Relying on self-generation makes sense for industrial firms because it is both less costly and more reliable. But it results in a loss in overall efficiency because captive generators, with a typical size of 1–2 megawatts, are much less fuel-efficient than utility-scale power plants.

Social Distortions

Social distortions occur when energy is priced below its socially efficient level. These types of distortions are most prominent in coal-based power generation, but they affect other types of fossil fuel consumption and mining as well. Natural gas is a cleaner type of fuel, but its inefficient allocation and use have increased reliance on oil-based power generation, a much more emission-intensive mode of production.

The environmental benefits of gas relative to coal can also be wiped out by supply chain leaks. When gas is leaked, it releases methane, an extremely potent greenhouse gas. Kerosene lamps and inefficient captive power generation are contributing to higher levels of indoor and outdoor air pollution with deleterious health effects. The highly subsidized electricity available to farmers has triggered groundwater depletion in India and parts of Pakistan.

UPSTREAM: UNPRICED EXTERNALITIES OF FOSSIL FUEL COMBUSTION AND COAL MINING

Failure to include the external costs of emissions in energy pricing results in excessive production of energy and emissions. Air pollution produced by fossil fuel–fired power plants, such as sulfur dioxide, nitrogen oxides, volatile organic compounds, and fine particulate matter, has been linked to various diseases and is a leading risk factor for mortality in South Asia. In India, exposure to inhalable particles with a diameter of 2.5 micrometers or less released from coal power plants contributes to an estimated 82,900 deaths in 2015 (Global Burden of Disease MAPs Working Group 2018). Ozone emissions, partially caused by nitrogen oxides emitted from gas-fired power plants, caused an estimated 5,000 deaths in Pakistan and 7,900 in Bangladesh in 2015 (Health Effects Institute 2017). Fossil fuel–based power generation is also the largest source of greenhouse gas emissions, contributing to 1.2 billion tons of carbon dioxide a year, or 44.5 percent of the regional total (IEA 2017a).

In India, every ton of coal combustion causes $190 in local health damage and $75 in climate change damage (Parry and others 2014). The average price of coal for power generation is $25 per ton. The marginal health and climate change damages associated with coal combustion thus significantly outweigh the private cost of coal in India. Local health damage accounts for the largest fraction of the social cost of coal consumption. Cutting excessive emissions from power generation is therefore not only good for the climate but also in the country's self-interest. In Bangladesh and Pakistan, the health damage associated with gas combustion is estimated at $0.21 and $0.17 per gigajoule (GJ), respectively, and the climate change damage is estimated at $2.41 per GJ (Parry and others 2014). The average price of domestic gas in Bangladesh and Pakistan is $1.40 and $2.30 per GJ, respectively.

Many remedies, such as emissions standards or mandatory adoption of abatement technologies, have been proposed to cut emissions. But economists have long established that market-based policy instruments—such as a Pigouvian tax, which raises the marginal private cost to the marginal social cost of energy—is the most cost-effective way to achieve a socially optimal level of emissions (Montgomery 1972). There is no environmental taxation of fossil fuel–fired power generation in Bangladesh and Pakistan. In 2010 India introduced a clean environmental cess (tax). In 2016 it was set at $6 per ton of coal. Although imposing this carbon tax is an important step in reducing environmental distortions, the level of the tax is only a fraction of the estimated marginal social cost of coal use in India.

Despite the higher cost of oil and diesel, Bangladesh and Pakistan have increased their use of liquid fuel for power generation as a way to cope with their shortfalls in domestic gas supply as of 2015 (IEA 2017b). Compared with gas, burning diesel and furnace oil emits a third more carbon dioxide and six times the amount of nitrogen oxide (Figure 1.9). Diesel also produces three times more emissions of methane, a potent climate warming agent, than does gas. Over the course of a century, methane will trap 28–34 times as much heat as an equivalent amount of carbon dioxide (IPCC 2013).

FIGURE 1.9 Diesel and fuel oil are much more polluting than natural gas

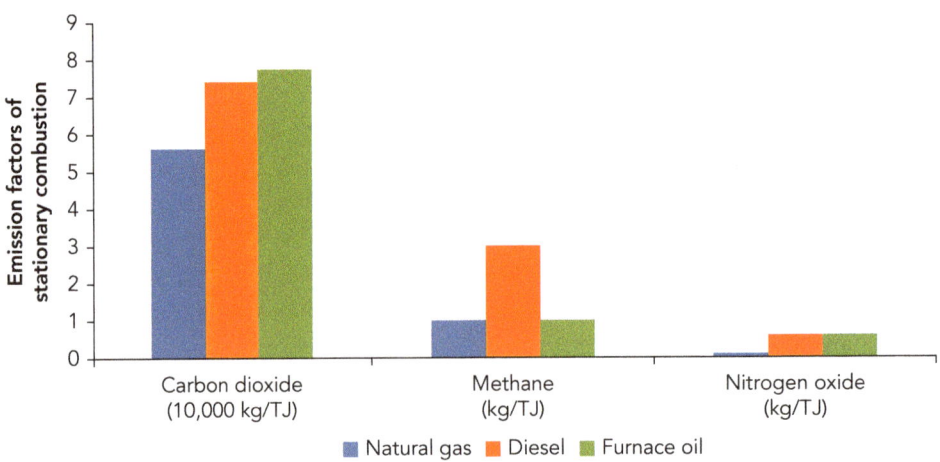

Source: IPCC (2006).
Note: kg/TJ = kilograms per terajoule.

Methane is also released during gas production, transmission, and distribution. If just 3 percent of gas is leaked into the air before combustion—a threshold that could be easily passed in Pakistan where loss of gas in the network reached about 14 percent in fiscal 2015—a gas plant does more climate damage than a coal plant (Alvarez and others 2012).

The mining and production of coal also impose large external costs. In India, for example, coal mining has caused fatalities, increased health risks for miners and mining communities, affected communities through displacement and resettlement, and resulted in ecological degradation. One study estimates that coal mining accounts for 22 percent of the total social costs of coal in the United States (Epstein and others 2011).

CORE: GROUNDWATER DEPLETION

Heavy subsidization of electricity tariffs for agriculture has led to overexploitation of groundwater for irrigation. Farmers in India pay a fixed amount based on the horsepower rating of the electric motor used rather than the metered consumption of power. Although some states such as Gujarat and West Bengal have adopted tariff reform, flat-rate tariffs remain in place in many states. Official government statistics indicate that 16 states have unmetered agriculture power supply and fixed tariff rates, and one state (Karnataka) provides electricity for agriculture free of charge (CEA 2014). At an average rate of $0.026 per kWh, the current tariffs for agriculture are much lower than the estimated cost of supply ($0.07 per kWh).

Under the current tariff structure, farmers have an incentive to overpump. India has in fact become the world's largest user of groundwater. Many areas have spawned active groundwater markets, where water is pumped not only for own-irrigation but also for resale to neighbors. Cheap irrigation has also influenced cropping patterns, prompting farmers to produce water-intensive crops, particularly rice. Between fiscal 1951 and fiscal 2014, the area irrigated by groundwater in India increased by almost seven times (Figure 1.10). Although many factors could contribute to increased groundwater extraction, such as changes in land use and rainfall pattern, the literature has established a causal link between electricity subsidies and groundwater overexploitation (Badiani and Jessoe 2013).

The current consumption of groundwater in India is unsustainable. According to the latest ground water resources assessment jointly carried out by Center Ground Water Board and State Ground Water Departments, 16 percent of the assessment units (block/taluks/mandals/watershed/firkka) are classified as over-exploited and 10 percent as semi-critical. Satellite-based analysis reveals that groundwater is being extracted more quickly than it is being replenished in northwestern and southern parts of the country. If the current trend continues, 60 percent of India's aquifers will be in critical condition by 2025 (Gulati and Pahuja n.d.). Because about 60 percent of India's agriculture depends on groundwater for irrigation and 85 percent of the rural population and 45 percent of urban population use it to meet drinking water needs, the depletion

FIGURE 1.10 **Groundwater extraction has surged in India**

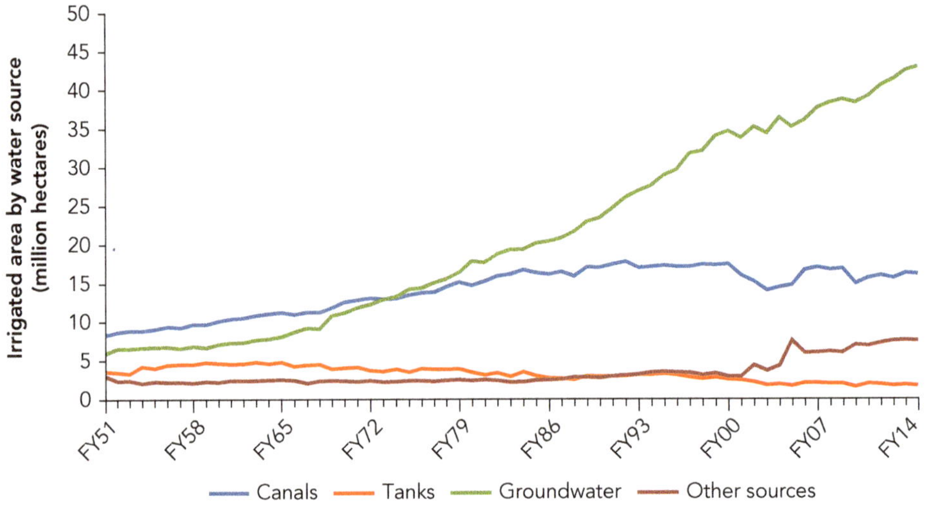

Source: Ministry of Agriculture and Farmers Welfare, India, accessed through Indiastat.
Note: FY = fiscal year.

of groundwater will have severe consequences for farmers' livelihood and for food and water security (Sekhri 2013).

DOWNSTREAM: DEPENDENCE ON KEROSENE LAMPS AND INEFFICIENT CAPTIVE GENERATORS

The kerosene or biomass fuels such as dung or wood used for residential lighting are known emitters of black carbon (Kurokawa and others 2013), which is released when fossil fuels, biofuels, and biomass are burned incompletely. Globally, black carbon is a major source of atmospheric warming, second only to carbon dioxide. It remains in the atmosphere for only a few days, but its impact on global warming during this short time is several hundred times greater than the same amount of carbon dioxide over 100 years (Jacobson and others 2013).

In South Asia, households and businesses without access to reliable electricity use almost 244 million kerosene lamps as a source of lighting (Tedsen 2013). Because of the heavy dependency on these lamps, the region has one of the world's highest rates of radiative forcing (atmospheric warming) effects of black carbon emissions from kerosene lighting (Map 1.1). Black carbon also contributes to glacial melting in the Himalayas and increases the disruption of South Asian monsoon patterns (Shindell and others 2012).

Kerosene lamps also pose significant health and safety risks. Kerosene is highly inflammable and often leads to accidental fires. In India, 2.5 million people a year suffer severe burns caused by overturned kerosene lamps (IFC 2010). Many studies also

MAP 1.1 Warming effects of black carbon emitted by kerosene lamps are greatest in South Asia

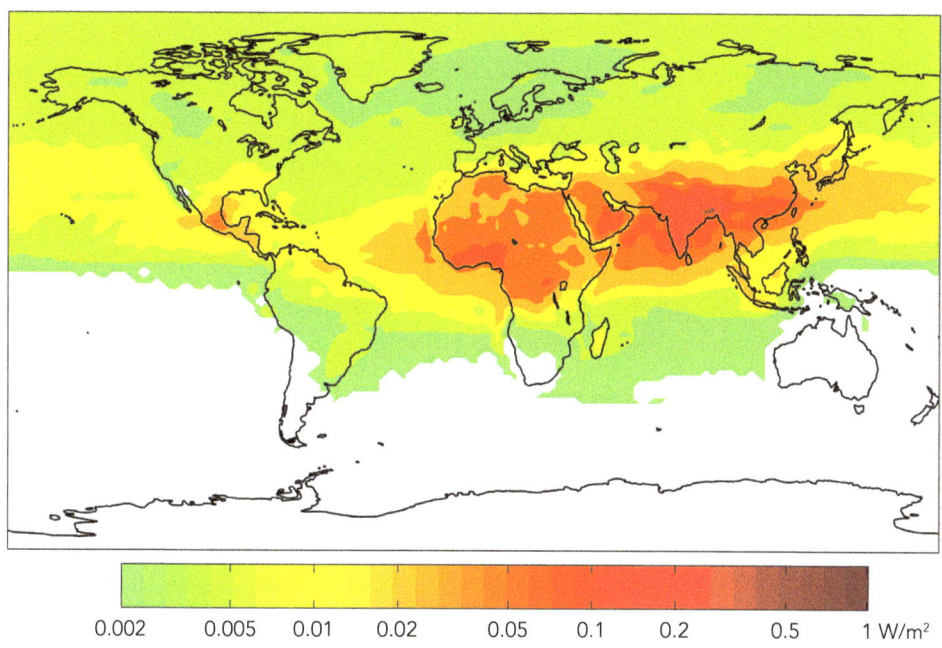

Source: Lam and others (2012).
Note: W/m² = watts per square meter.

report a strong association between exposure to kerosene lamps and tuberculosis, respiratory infections, and lung diseases because almost all emissions from kerosene lamps are minuscule particles that can be inhaled deep into the lungs (Apple and others 2010; Sahu and others 2011).

Self-generation of power is gaining importance among businesses, as firms seek sources of power supply other than the grid. According to the latest World Bank Enterprise Surveys, 63 percent of firms in Bangladesh, 47 percent in India, and 65 percent in Pakistan have some kind of backup generator. These shares are substantially higher than the average, 39 percent, in lower-middle-income countries.

"Captive" power generators are both wasteful and polluting. They are less efficient than utility-scale power plants because they are smaller and have lower use rates (they are used only as backups). In Pakistan, for example, the average efficiency of captive generators is estimated at 18–28 percent—far lower than the 35 percent of thermal power plants (Bhutta 2015). Pollution from captive plants is also two to three times more damaging to human health than pollution from large power plants because, unlike large power plants that have high stacks and are located far away from population centers, captive generators are closer to where people live and are typically not equipped with pollution control equipment (WHO 2015).

References

Abeberese, A. B. 2017. "Electricity Cost and Firm Performance: Evidence from India." *Review of Economics and Statistics* 99 (5): 839–52.

ADB (Asian Development Bank). 2014. *Industrial Energy Efficiency Opportunities and Challenges in Bangladesh*. Manila, Philippines.

Allcott, H., A. Collard-Wexler, and S. D. O'Connell. 2016. "How Do Electricity Shortages Affect Industry? Evidence from India." *American Economic Review* 106 (3): 587–624.

Apple, J., R. Vicente, A. Yarberry, N. Lohse, E. Mills, A. Jacobson, and D. Poppendieck. 2010. "Characterization of Particulate Matter Size Distributions and Indoor Concentrations from Kerosene and Diesel Lamps." *Indoor Air* 20: 399–411.

Alvarez, Ramón A., Stephen W. Pacala, James J. Winebrake, William L. Chameides, and Steven P. Hamburg. 2012. "Greater Focus Needed on Methane Leakage from Natural Gas Infrastructure." *PNAS* 109 (17): 6435–40.

Badiani, Reena, and Katrina Jessoe. 2013. "The Impact of Electricity Subsidies on Groundwater Extraction and Agricultural Production." Working Paper, University of California, Davis.

Banerjee, Sudeshna Ghosh, Douglas Barnes, Bipul Singh, Kristy Mayer, and Hussain Samad. 2015. *Power for All: Electricity Access Challenge in India*. Washington, DC: World Bank.

Bhutta, Zafar. 2015. "No Fuel Oil: New Power Plants to Run on Coal or LNG." Express Tribune, February 5. https://tribune.com.pk/story/833620/no-fuel-oil-new-power-plants-to-run-on-coal-or-lng.

BPDB (Bangladesh Power Development Board). Various years. *Annual Report*. Dhaka.

———. 2015. "Retail Tariff Rate of BPDB." Database. http://www.bpdb.gov.bd/bpdb/index.php?option=com_content&view=article&id=231.

CEA (Central Electricity Authority). 2014. "Tariff and Duty of Electricity Supply in India." New Delhi, March. http://www.cea.nic.in/reports/others/enc/fsa/tariff_2014.pdf.

———. 2015. "Daily Coal Reports" (database). http://www.cea.nic.in/dailycoal.html.

———. 2018. *Load Generation Balance Report 2017–18*. New Delhi.

CERC (Central Electricity Regulatory Commission). 2015. *Annual Report 2014–2015*. New Delhi.

———. 2017. *Report on Short-Term Power Market in India: 2016–17*. Economics Division, New Delhi.

CIL (Coal India Ltd.) 2016. *Annual Report*. New Delhi.

EIA (Energy Information Administration). 2015. "India's Coal Industry in Flux as Government Sets Ambitious Coal Production Targets." *Today in Energy*. https://www.eia.gov/todayinenergy/detail.php?id=22652.

Epstein, Paul.R., Jonathan J. Buonocore, Kevin Eckerle, Michael Hendryx, Benjamin M. Stout III, Richard Heinberg, Richard W. Clapp, Beverly May, Nancy L. Reinhart, Melissa M. Ahern, Samir K. Doshis, and Leslie Glustrom. 2011. "Full Cost Accounting for the Life Cycle of Coal." *Annals of New York Academy of Sciences* 1219: 73–98.

Fisher-Vanden, K., E. T. Mansur, and Q. J. Wang. 2015. "Electricity Shortages and Firm Productivity: Evidence from China's Industrial Firms." *Journal of Development Economics* 114: 172–88.

Global Burden of Disease MAPs Working Group. 2018. *Burden of Disease Attributable to Major Air Pollution Source in India Special Report*. Boston.

Government of India. 2015. Economic Survey of India. New Delhi.

Grainger, C. A., and C. D. Kolstad. 2010. "Distribution and Climate Change Policies." In *Climate Change Policies: Global Challenges and Future Prospects*, edited by E. Cerda and X. Labandiera. Cheltenham, U.K.: Edward Elgar Publishing.

Grainger, C.A., and Fan Zhang. 2017a. "The Impact of Electricity Shortages on Firms: Evidence from Pakistan." Policy Research Working Paper, WPS8130, World Bank, Washington, DC.

Grainger, C. A., and Fan Zhang. 2017b. "The Impact of Electricity Shortages on Micro- and Small-Enterprises: Evidence from India." Background paper prepared for this report, World Bank, Washington, DC.

Grainger, Corbett, Fan Zhang, and Andrew Shreiber. 2015. "Distributional Impact of Energy Cross-Subsidies in Transition Economies: Evidence from Belarus." Policy Research Working Paper 7385, World Bank, Washington, DC.

Gulati, Mohinder, and Sanjay Pahuja. n.d. "Direct Delivery of Power Subsidy to Agriculture in India." Energy Sector Management Assistance Program (ESMAP), World Bank, Washington, DC.

Health Effects Institute. 2017. *State of Global Air: A Special Report on Global Exposure to Air Pollution and Its Disease Burden*. Boston.

Hsieh, C.-T., and P. J. Klenow. 2014. "The Life Cycle of Plants in India and Mexico." *Quarterly Journal of Economics* 129 (3): 1035–84.

Hydrocarbon Development Institute of Pakistan. Various years. *Pakistan Energy Yearbook*. Islamabad.

IEA (International Energy Agency) 2017a. CO_2 Emission from Fuel Combustion 2017 (database). Paris. http://data.iea.org/payment/products/115-co2-emissions-from-fuel-combustion-2017-edition.aspx.

———. 2017b. World Energy Statistics and Balances (database). Paris. https://www.iea.org/statistics/relateddatabases/worldenergystatisticsandbalances/

IFC (International Finance Corporation). 2010. *Solar Lighting for the Base of the Pyramid: Overview of an Emerging Market*. Washington, DC.

Indian Planning Commission. Various years. *On the Working of State Power Utilities and Electricity Departments*. New Delhi.

IPCC (Intergovernmental Panel on Climate Change). 2006. *Guidelines for National Greenhouse Gas Inventories*. Geneva.

———. 2013. "Working Group I Contribution to the IPCC Fifth Assessment Report Climate Change 2013: The Physical Science Basis. Final Draft Underlying Scientific-Technical Assessment." Geneva.

Jacobson, Arne, Tami Bond, Nicholas L. Lam, and Nathan Hultman. 2013. *Black Carbon and Kerosene Lighting: An Opportunity for Rapid Action on Climate Change and Clean Energy for Development*. Policy Paper 2013-03, Brookings Institution, Washington, DC.

Kerkhof, A. C., Henri C. Moll, Eric Drissen, and Harry C. Wilting, 2008. "Taxation of Multiple Greenhouse Gases and the Effects on Income Distribution: A Case Study of the Netherlands." *Ecological Economics* 67: 318–26.

Kornai, János. 1979. "Resource-Constrained versus Demand-Constrained Systems." *Econometrica* 47 (4): 801–19.

Kurokawa, J., T. Ohara, T. Morikawa, S. Hanayama, G. Janssens-Maenhout, T. Fukui, K. Kawashima, and H. Akimoto. 2013. "Emissions of Air Pollutants and Greenhouse Gases over Asian Regions during 2000–2008: Regional Emission Inventory in Asia (REAS) Version 2." *Atmospheric Chemistry and Physics* 13 (21): 11019–58.

Laffont, Jean-Jacques, and Jean Tirole. 1993. *A Theory of Incentives in Procurement and Regulation.* Cambridge, MA: MIT Press.

Lam, Nicholas L., Yanju Chen, Cheryl Weyant, Chandra Venkataraman, Pankaj Sadavarte, Michael A. Johnson, Kirk R. Smith, Benjamin T. Brem, Joseph Arineitwe, Justin E. Ellis, and Tami C. Bond. 2012. "Household Light Makes Global Heat: High Black Carbon Emissions from Kerosene Wick Lamps." *Environmental Science and Technology* 46 (24): 13531–38.

Li, Y., and M. Rama. 2015. "Firm Dynamics, Productivity Growth, and Job Creation in Developing Countries: The Role of Micro- and Small Enterprises." *World Bank Research Observer* 30 (1): 3–38.

McRae, S. 2015. "Infrastructure Quality and the Subsidy Trap." *American Economic Review* 105 (1): 35–66.

Montgomery, W. David. 1972. "Markets in Licenses and Efficient Pollution Control Programs." *Journal of Economic Theory* 5 (3): 395–418.

NEPRA (National Electric Power Regulatory Authority). Various years. *State of Industry Report.* Islamabad.

Ministry of Coal, India. 2017. *Annual Report.* New Delhi.

———. Various years. *Provisional Coal Statistics.* New Delhi.

Ministry of Finance, Bangladesh. 2015. *Bangladesh Economic Review.* Dhaka.

Ministry of Finance, India. 2017. *Indian Economic Survey.* New Delhi.

Ministry of Finance, Pakistan. 2017. *Pakistan Economic Survey 2016–2017.* Islamabad.

National Transmission and Distribution Center. 2014. *Electricity Demand Forecast Based on Multiple Regression.* Government of Pakistan, Islamabad.

Parry, Ian, Dirk Heine, Eliza Lis, and Shanjun Li. 2014. *Getting Energy Prices Right: From Principle to Practice.* Washington, DC: International Monetary Fund.

Petrobangla. 2015. *Annual Report 2015.* Dhaka. https://petrobangla.org.bd/?params=en /annualreport.

———. 2016. *Annual Report 2016.* Dhaka. https://petrobangla.org.bd/?params=en/annualreport.

Rahman, M. Azizur. 2015. "Chevron Abandons $650m Investment Plan." January 14. http://today .thefinancialexpress.com.bd/public/first-page/chevron-abandons-$650m-investment-plan.

Sahu, M., J. Peipert, V. Singhal, G. N. Yadama, and P. Biswas. 2011. "Evaluation of Mass and Surface Area Concentration of Particle Emissions and Development of Emissions Indices for Cookstoves in Rural India." *Environmental Science and Technology* 45: 2428–34.

Samad, Hussain, and Fan Zhang. 2016. "Benefits of Electrification and the Role of Reliability: Evidence from India." Policy Research Working Paper 7889, World Bank, Washington, DC.

———. 2017. "Heterogeneous Effects of Rural Electrification: Evidence from Bangladesh." Policy Research Working Paper 8102, World Bank, Washington, DC.

———. 2018. "Electrification and Women Empowerment: Evidence from Rural India." Policy Research Working Paper, World Bank, Washington DC.

Schwab, Klaus. 2018. *The Global Competitiveness Report 2017–2018*. Geneva: World Economic Forum.

Sekhri, S. 2013. "Missing Water: Agricultural Stress and Adaptation Strategies in Response to Groundwater Depletion among Farmers in India." Working Paper, International Growth Centre, London School of Economics and Political Science.

Shindell, Drew, Johan C. I. Kuylenstierna, Elisabetta Vignati, Rita van Dingenen, Markus Amann, Zbigniew Klimont, Susan C. Anenberg, Nicholas Muller, Greet Janssens-Maenhout, Frank Raes, Joel Schwartz, Greg Faluvegi, Luca Pozzoli, Kaarle Kupiainen, Lena Höglund-Isaksson, Lisa Emberson, David Streets, V. Ramanathan, Kevin Hicks, N. T. Kim Oanh, George Milly, Martin Williams, Volodymyr Demkine, and David Fowler. 2012. "Simultaneously Mitigating Near-Term Climate Change and Improving Human Health and Food Security." *Science* 335 (6065): 183–89.

Shleifer, Andrei, and Robert Vishny. 1994. "Politicians and Firms." *Quarterly Journal of Economics* 109: 995–1025.

Tedsen, E. 2013. "Black Carbon Emissions from Kerosene Lamps: Potential for New CCAC Initiative." Ecological Institute, Berlin.

USAID (U.S. Agency for International Development). 2011. *Evaluation of Economic Value of Natural Gas in Various Sectors.* Washington, DC.

———. 2013. *The Causes and Impacts of Power Sector Circular Debt in Pakistan.* Study commissioned by the Planning Commission of Pakistan. Washington, DC.

World Bank. Various years. Global Economic Monitor Commodities (database). Washington, DC. http://databank.worldbank.org/data/reports.aspx?source=global-economic-monitor -commodities.

———. 2015a. "Pakistan—Second Power Sector Reform Development Policy Credit Program Project Document." Washington, DC.

———. 2015b. "Pakistan Country Snapshot (2015)." Washington, DC. October.

———. 2015c. "A Review of Renewable Investment and Power System Operational Issues in Bangladesh." Subsequently published in 2017 in *Renewable and Sustainable Energy Reviews* 650–58.

———. 2018. World Development Indicators (database). Washington, DC. http://databank .worldbank.org/data/reports.aspx?source=world-development-indicators.

WHO (World Health Organization). 2015. *Reducing Global Health Risks through Mitigation of Short-Lived Climate Pollutants: Scoping Report for Policymakers.* Geneva.

CHAPTER 2

Assessing the Cost of Distortions

The power sector in South Asia is riddled with externalities and policy-induced distortions, as described in Chapter 1. To disentangle the impact of distortions, the analysis decomposes the overall cost of distortions into three partial costs: institutional, regulatory, and social. Each partial cost is defined as the difference in welfare between a partial equilibrium and the baseline. The baseline is the actual equilibrium, characterized by multiple distortions (Table 2.1).

The partial cost is estimated by removing distortions, one at a time. In this way it is possible to estimate the cost associated with that distortion through comparison with the actual equilibrium. In estimating institutional cost, consider a thought experiment that asks what would happen if resources were allocated efficiently in the input and output markets of electricity and if public utilities were as efficient as their private peers in supplying power. All else being equal, the supply of energy would increase and shortages would decline. Under this "what if" scenario, called the *efficient equilibrium*, consumers would gain from extra energy consumption and producers would gain from reduced production costs. The cost of the institutional distortion is the combined loss for consumers and producers—that is, the deadweight loss to the economy from departing from this efficient equilibrium.

To estimate regulatory cost, the analysis asks what if energy prices were no longer regulated but were instead set equal to the private cost of supply. Consumers would respond to higher prices by reducing demand, and producers would increase production until the market cleared. Under this *market equilibrium*, consumers would benefit from elimination of the unmet demand for energy, and producers would benefit from the expansion of profitable sales. The cost of the regulatory distortion is the deadweight loss to the economy from departing from this market equilibrium, holding all else equal.

TABLE 2.1 Decomposition of the cost of distortions

Type of equilibrium	Institutional distortion	Regulatory distortion	Social distortion	Cost decomposition
Actual	Yes	Yes	Yes	Baseline = Output A
Efficient	No	Yes	Yes	Institutional cost = Output E—Output A
Market	Yes	No	Yes	Regulatory cost = Output M—Output A
Social	Yes	Yes	No	Social cost = Output S—Output A

Note: Welfare is loosely referred to as output in the table.

The social cost is defined as the deadweight loss from pricing energy below the overall cost to society of its consumption. Imposing social pricing would increase tax revenue and reduce excessive energy consumption (and the emissions and health damages associated with it). But consumers and producers would also be adversely affected by the higher price and lower quantity. The social cost is calculated as the net changes in budgetary revenue, environmental externalities, and consumer and producer surpluses from departing from the social equilibrium.

The size of any distortion, while conceptually easy to specify, depends crucially on two main elements. One is the shape of the demand and supply curves for energy—that is, how sensitive energy producers and consumers are to price changes. The other is a set of parameters such as the monetary value of the health effects of pollution and the exact efficiency gap between public and private utilities, which determine how the supply and demand curve would shift when distortions are removed. Historical data on price and quantity are collected to determine the price sensitivity of producers and consumers in the fuel and electricity markets. Disaggregated data on utilities, households, and firms are collected to estimate the values of the parameters.

This chapter first presents graphic representations of the three partial costs. It then explains how to estimate key parameters in order to determine each counterfactual equilibrium and describes the data sources used.

Decomposing the Cost of Distortions

This section uses graphic examples to illustrate the economic framework for quantifying the costs of distortions. Figures 2.1–2.3 use supply and demand diagrams for energy (fuels and electricity) to show the effects of distortions on consumer and producer well-being—that is, consumer and producer surplus (Box 2.1).

BOX 2.1 The welfare economics of energy

In economics, the quantity of a good consumed depends on its demand and supply curves. Each point on the demand curve represents the price consumers are willing to pay for a given quantity, and each point on the supply curve represents the amount producers are willing to supply at a given price, all else being equal. Market equilibrium occurs when these curves intersect—that is, when demand equals supply. In the context of the energy market, Figure B2.1.1 shows that, at the market-clearing price of P_M per unit, Q_M units of energy will be supplied and consumed.

For all quantities up to Q_M, consumers are willing to pay a rate higher than the market-clearing price. For example, at Q_L units of energy, consumers are willing to pay P_C per unit—more than the market price of P_M per unit. The difference between the price consumers are willing to pay and the price they actually pay is the private benefit or surplus reaped by the consumer. This benefit, called the *consumer surplus*, is depicted as the area below the demand curve but above the market-clearing price.

FIGURE B2.1.1 Price at market-clearing level

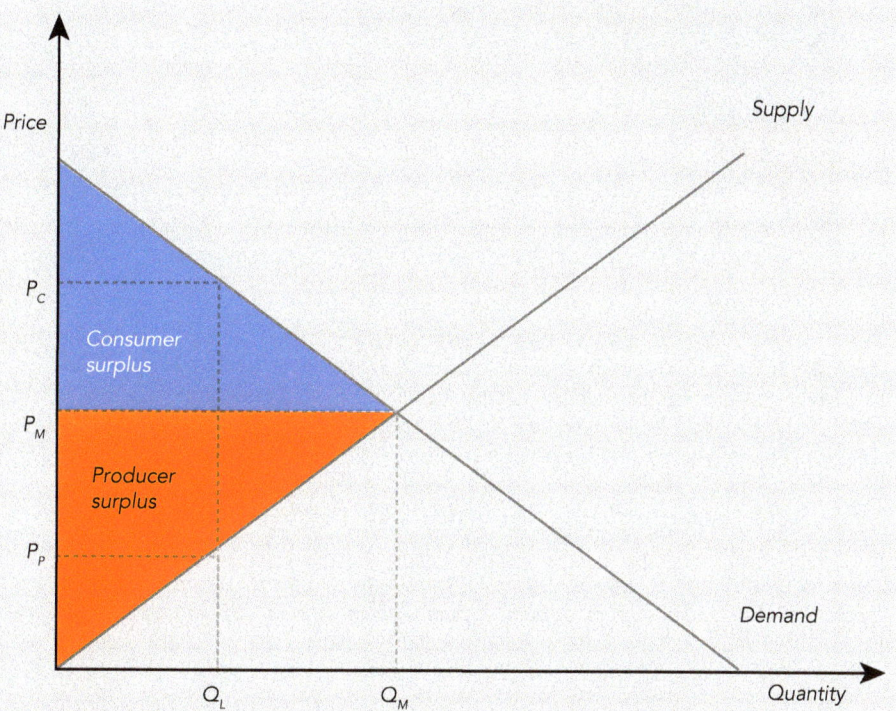

box continues next page

BOX 2.1 **The welfare economics of energy** (continued)

FIGURE B2.1.2 Welfare loss from underpricing

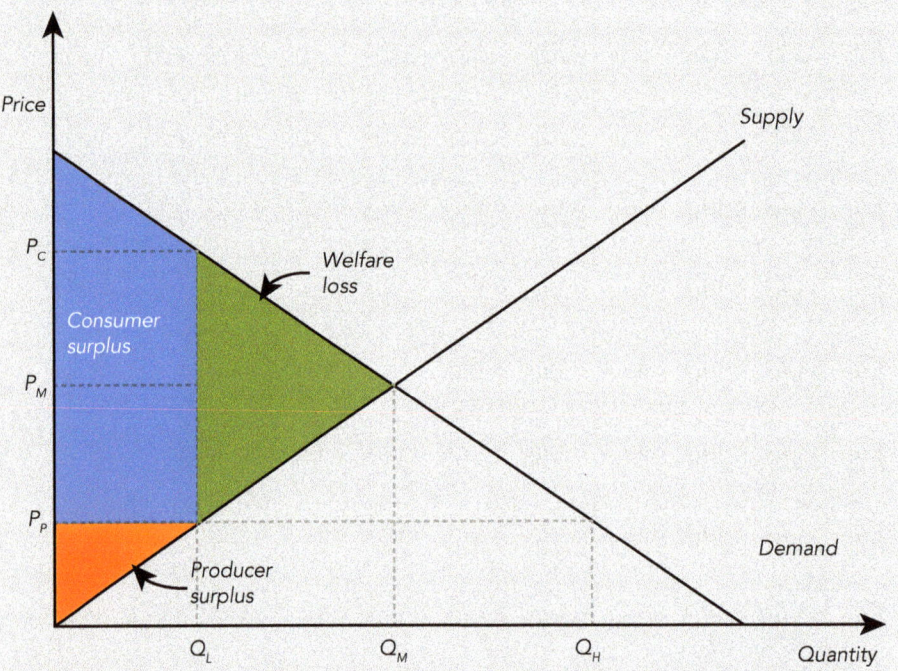

Producers are willing to accept a payment of P_P per unit if they supply Q_L units of energy. But, because the market-clearing price is higher than P_P per unit, producers receive more than the minimum price they are willing to accept until they supply Q_M units of energy. This difference is called the *producer surplus* because it represents the additional benefit reaped by the producer. It is depicted as the area above the supply curve but below the market-clearing price. The combined producer and consumer surplus represents the overall economic benefit or surplus reaped by producers and consumers by interacting in a free market with no price controls or quotas.

Consider an alternative scenario in which the price is administratively set at P_P per unit, below the market price of P_M per unit (Figure B2.1.2). At this lower price, consumers will demand a higher amount of energy, Q_H, but producers will be willing to supply only Q_L units, less than the market-clearing quantity of Q_M units. As a result, only Q_L units will be traded in the market, reducing the net surplus or benefit that can be attained by producers and consumers. This welfare loss is called a *deadweight loss*.

INSTITUTIONAL COST

To illustrate how to quantify each of the partial cost, the rest of the chapter uses the coal market for power generation in India as an example. The solid lines in Figure 2.1 show the actual supply and demand for coal. The dashed lines show the counterfactual supply corresponding to efficient production. For purposes of illustration, the discussion here does not consider the import of coal (the world supply of coal or gas is considered in the actual estimations). The actual equilibrium is therefore at the intersection of the admin-istered price and the domestic supply of coal (Q_A). At the prevailing price, the quantity desired by consumers is much larger than the amount producers are willing to provide. The difference is considered the coal shortage or unmet demand.

If institutional distortions were removed—by, for example, introducing effec-tive competition in the coal market, which would stimulate technology upgrades and innovation—firms would produce more coal with the same amount of variable inputs. The supply curve would thus shift to the right. The new equilibrium point, the *efficient equilibrium*, would occur at the intersection of the original administered price and the efficient supply curve. Efficient output (Q_E) would be greater than actual output because with a lower marginal production cost firms would be willing to produce more at any given price than they did before. Increased production would lead to a gain in pro-ducer surplus, represented by area B. Consumers would also benefit from the increase in output because at the efficient equilibrium unmet demand would be reduced by the amount of $Q_E - Q_A$. Unmet demand means there are consumers who are willing to pay

FIGURE 2.1 Estimating the welfare cost of institutional distortion

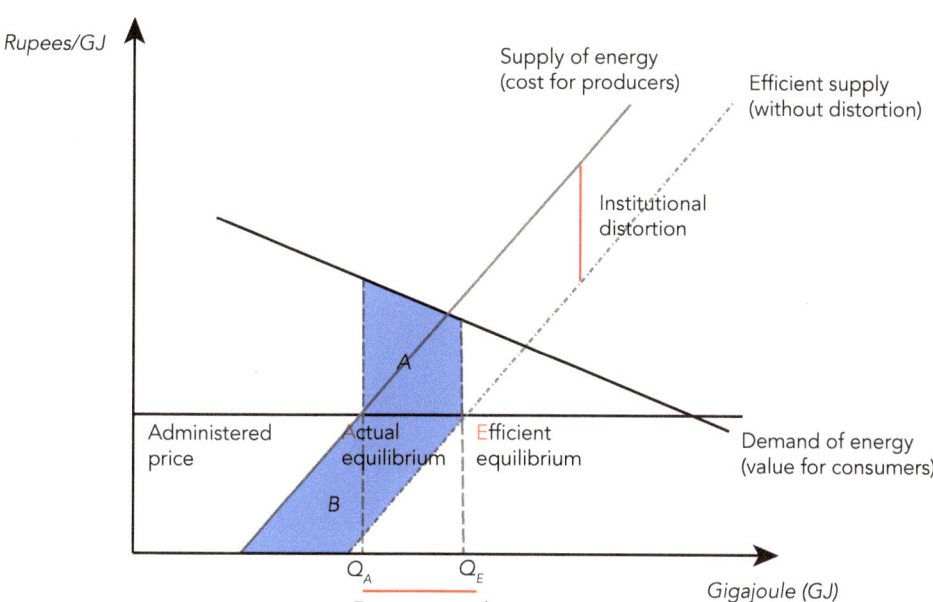

higher than the prevailing price for energy. If their unmet demand is served, these consumers are made better off. The resulting increase in consumer surplus is represented by area A. The overall institutional cost, linked to forgone supply $(Q_E - Q_A)$, is the sum of areas A and B—that is, the gains in consumer and producer surplus that are forgone under the institutional distortions.

REGULATORY COST

Figure 2.2 illustrates the effects of regulatory distortions on market outcomes—here in the form of an administered price. If regulatory distortions were removed, there would be no shift in the supply curve, but the now-liberalized price would be fully determined by market forces. The price would rise, causing demand to decrease and supply to expand. The market would clear when supply equals demand. The new equilibrium, the *market equilibrium*, would be at a higher price and a larger quantity than before.

Compared with the market equilibrium, the amount desired by consumers at the administered price (Q_D) is well above the level of consumption that would be obtained at the market price (Q_M). The price ceiling also artificially depresses output. Allowing the market to determine the price and quantity would thus increase the size of the economic pie.

Under the market equilibrium, consumers who were previously charged the administered price would be worse off because they now have to pay a higher price. Consumers who were willing to pay more for energy than the administered price but

FIGURE 2.2 **Estimating the welfare cost of regulatory distortion**

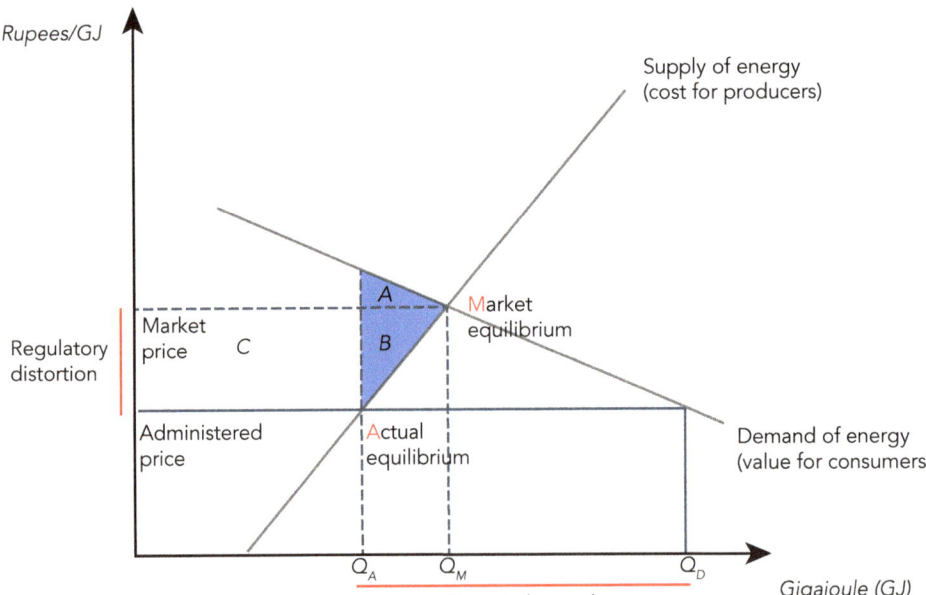

experienced shortages would be better off because their unmet demand for energy would now be served. The overall change in consumer surplus is the area A – C. Producers always benefit from price liberalization because the larger quantity and higher price mean higher profits. The change in producer surplus is the area C + B. The net regulatory cost caused by deviation from market-driven pricing is thus represented by the blue triangle (A + B), which consists of both the gains in consumer surplus (area A) and the gains in producer surplus (area B) that are forgone under the administered price.

SOCIAL COST

The previous scenarios consider only the private cost of energy supply. However, there is also a social cost because energy consumption generates side effects, such as emissions of toxic local pollutants and greenhouse gases, that cause undue harm to others. These effects are called *externalities*. Unpriced externalities introduce inefficiencies because they create waste in the form of excessive external costs.

To achieve socially optimal pricing, an environmental tax equal to the marginal damage of emissions could be introduced to internalize the cost of externalities, as shown in Figure 2.3. The tax makes energy production more expensive, shifting the private supply curve to the left to match the social supply curve.

In the spirit of partial equilibrium, assume that all other factors remain the same and the consumer price is still the original government-administered price. At the

FIGURE 2.3 Estimating the welfare cost of social distortion

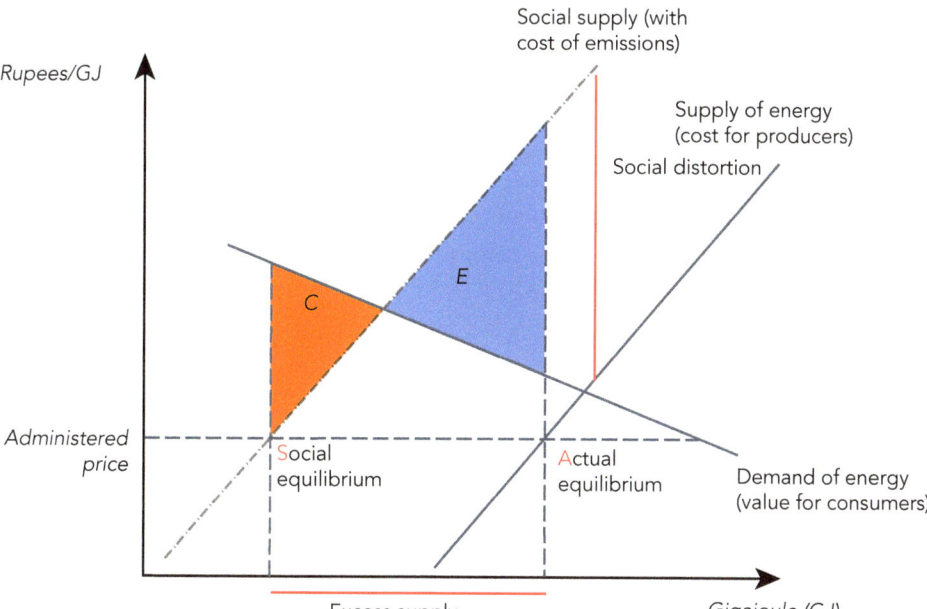

new equilibrium, called the *social equilibrium*, output is smaller, reducing emissions and other externalities. The tax would also generate additional government revenue that could be recycled back into the economy. However, because consumption is reduced and producers have to pay a tax, there is also a loss in consumer and producer surplus. The overall social cost caused by deviation from social pricing consists of the net change in external cost, tax revenue, consumer surplus, and producer surplus, represented by E – C in the figure. Removing social distortions always produces greater benefits than costs because the avoided harmful external costs (E) are substantially larger than the surplus loss from reduced consumption (C).

Estimating the Cost of Distortions

Constructing the counterfactual equilibria illustrated in the previous section requires several parameters. To predict market equilibrium, one needs to know how sensitive energy production and consumption are to a change in price (the *price elasticities of supply and demand*). If producers and consumers do not respond to price changes, a price intervention will have no impact on the market equilibrium. The more responsive they are—that is, the greater the elasticity of supply and demand—the greater will be the economic waste caused by distortions.

In the short run, supply and demand for energy are both relatively inelastic. But in the long run there is a great deal of scope for both producers and consumers to respond to price adjustments. Using historical price and quantity data, the analysis estimates the long-run elasticities of supply and demand for energy, which are used to identify the changes in producer and consumer surplus associated with a departure from counterfactual equilibria.

Identifying an efficient equilibrium also requires estimating the new supply curve. A rise in productivity or a reallocation of resources from less efficient to more efficient units would increase supply and shift the supply curve to the right. To estimate the potential increase in aggregate supply following the removal of institutional distortions, the analysis calculates the distance from the production possibility frontier of utilities. It also simulates the relationship between fuel allocation and systemwide electricity output based on hourly dispatch models.

To estimate the social equilibrium, the analysis calculates the monetary value of environmental and health damages. For the marginal cost of greenhouse gas emissions, it uses the low-bound value for the shadow price of carbon dioxide of $40 per ton recommended by the World Bank Sustainable Development Practice Group. For the marginal cost of local pollutants, it relies on estimates by Parry and others (2014) of how much pollution from fossil fuel–based power plants is inhaled by the exposed population, how it contributes to illness given demographic characteristics, and the monetary equivalent of the health effects based on the value of a statistical life.

Because energy shortages are widespread in South Asia, the observed demand does not reflect the unexpressed latent demand for electricity among millions of unserved and underserved consumers. Moreover, in the absence of outages, firms would likely behave very differently—for example, producing at different times of the day or employing different mixes of capital and labor. For these reasons, the welfare effects of lack of reliable access to electricity on households and firms are complex and cannot be captured by a static supply and demand model. To quantify the impact of institutional distortions downstream, the analysis uses econometric methods to estimate the effects of power shortages on households and firms.

ESTIMATING SUPPLY AND DEMAND

To estimate the supply and demand for energy, the analysis assumes a constant elasticity of the supply and demand function. The log-linear relationship between price and quality is a standard assumption in the empirical literature (see Clements and others 2013; Coady and others 2017; Dahl 2006; Davis 2014),

$$Q = AP^\varepsilon,$$

where Q is quantity, P is price, ε is elasticity, and A is a scale factor. The analysis uses long historical time-series data on price and quantity to estimate elasticity. It then backs out the scale parameter for the demand and supply using currently observed prices and quantities.

Estimating the price elasticities of supply and demand is not straightforward for at least two reasons. First, price and quantity usually trend over time. Careful discernment of the presence of unit root and cointegration is needed to avoid a spurious regression in which the regression may be picking up a relationship between the trends in two variables rather than an underlying relationship between them. Second, price and quantity are simultaneously determined. With a finite sample, exogenous variations in price are needed to identify the coefficients. Third, energy supply and demand generally adjust slowly to changes in prices. Therefore, it is important to distinguish between short-run and long-run elasticities.

This report provides new measures of the price elasticities of the supply and demand for fuel (coal or gas) and electricity for Bangladesh, India, and Pakistan. The estimation approach uses an autoregressive distributed lag model to test cointegration. It uses lagged own prices, the lagged price of the imported substituting fuel, or the price of inputs (such as the wages of mine workers) to address concerns about endogeneity and serial correlation. It relies on generalized methods of moments to estimate an error correction model to obtain long-run price elasticities (see Appendix A for details on the methodology.) Comparison of the estimated elasticities with previous empirical estimates for South Asia or for countries in other regions confirms that they are within the range of those suggested in the literature.

CONSTRUCTING A PRODUCTION POSSIBILITY FRONTIER

The analysis uses different benchmarks to estimate institutional distortions reflected as a deviation from the efficient benchmark, or *production possibility frontier.* For the generation sector in Bangladesh, India, and Pakistan, private power plants are a performance benchmark for government-owned power plants. Using domestic power plants as a benchmark can control for most of the potentially confounding variations in plant performance that are common in cross-country comparisons such as fuel quality, market environment, and climate conditions.

Depending on data availability, the analysis estimates an input demand function (partial productivity measure), a production function (total factor productivity measure), or both. Each measure controls for a rich set of plant characteristics such as age, technology, the heat content of fuel inputs, and the plant's ownership type. For the partial productivity measure, the estimation focuses on the thermal efficiency of fuel-intensive power plants. For the total factor productivity measure, the analysis estimates a production stochastic frontier, which considers deviation from the frontier as caused by both technical inefficiency and random noise (see Appendix B for more details on the methodology).

The productivity literature has long recognized input endogeneity concerns associated with the fact that input and output decisions are determined simultaneously. Adding to the complexity of analysis for electricity generation, a plant's efficiency is heavily influenced by how often it is shut down or operated below capacity. Shutdowns and underutilization may reflect inefficiency, but they can also result from dispatch that does not always follow merit order. The analysis uses exogenous variations in dispatch caused by regionwide demand shocks to address both endogeneity concerns.

For the distribution sectors in Bangladesh, India, and Pakistan, the most efficient distribution utility in each country is the benchmark. The analysis uses the stochastic production frontier approach to estimate the technical inefficiency of individual distribution companies after controlling for differences in fixed assets, employee expenditure, peak demand, and consumer mix. This approach produces a more conservative estimation because even the most efficient domestic distribution utility may still be inefficient when compared to international norms.

SIMULATING THE INCREASE IN OUTPUT

Increasing generation efficiency and reallocating input fuel (gas) from less productive plants to more productive plants can lead to several important benefits. First, it can increase total power production and therefore reduce power cuts. Second, it can reduce total production costs as less expensive plants replace more expensive ones. Third, by increasing efficiency in the use of gas, it can allow gas-based generation to offset output from furnace oil– or diesel-based generation and thus reduce emissions.

But gains in generation and allocative efficiency do not translate linearly into gains in total output. The size of the overall benefits also depends on demand side considerations.

Because demand varies significantly between peak and nonpeak hours, the analysis first constructs hourly generation and demand profiles based on observed historical data. It then calculates the plant-level output increase, assuming that technical and allocative inefficiency is removed. Applying the efficiency improvement at an hourly resolution, plants first use the additional power generation to reduce unserved energy demand before offsetting production from more expensive plants. Emissions savings are calculated on the basis of emission factors and the output of individual plants.

ESTIMATING THE WELFARE EFFECTS ON HOUSEHOLDS AND FIRMS

To quantify the adverse effects of power shortages on households, the analysis compares changes in social and economic outcomes for households following grid connection with outcomes for households that remain off the grid. It uses nationally representative household survey data for all three countries. Gaining access to the grid is important for a broad range of social and economic benefits, but access is not enough: electrification yields full benefits only if a connected household receives adequate service. The analysis quantifies the value of a better-quality electricity supply by taking advantage of the variation in the reliability of electricity supply observed in the data.

Identifying the causal relationship between electrification and household welfare is not straightforward because grid expansion and a household's decision to adopt electricity may not be random. For example, the government may target for electrification projects areas that are more easily accessible and have greater potential for growth. Meanwhile, when electricity becomes available in a village, more well-off households are more likely to adopt it first. These preexisting differences between grid-connected and off-grid households may contribute to differential trends in income growth even in the absence of electrification. The analysis uses a two-stage propensity score–weighted fixed-effects model and instrumental variables to deal with the potential selection bias.

To quantify the effects of power shortages on firm-level productivity, the analysis matches industry survey data with power shortage data reported by utilities. It correlates firm-level output or value added with power shortages, while controlling for a firm's sector, and the costs of other factor inputs. Correlation does not mean causation, however, because power shortages are likely to be endogenous. The endogeneity could arise from the sorting of firms across locations. For example, more productive firms may be able to choose locations with better infrastructure facilities. In addition, the quality of the power supply is likely to be endogenous to output and growth: in regions with faster economic growth, the demand for electricity would be higher, which could in turn result in worse outages. In both cases, the effect of power outages would be underestimated. State policies could also affect both infrastructure spending and the business environment. The analysis uses a fixed-effects model and instrumental variable approach to address potential bias from the simultaneous causality.

Partial Equilibrium Analysis

Interactions between distortions are likely. The sum of three partial costs is therefore not the same as the full cost. It could be smaller or larger than the full cost, depending on whether interactions between distortions reinforce or offset one other. Consider an example. Excessive electricity consumption is associated with social distortions that lead to a large partial cost. An insufficient electricity supply caused by institutional distortions also leads to a large partial cost. The full cost in this example is smaller than the sum of the two partial costs because the two distortions (imperfectly) offset each other. The partial equilibrium analysis ignores second-best considerations (in the sense that it does not consider interactions between distortions), but it does offer a clean way to highlight issues in the power sector. Chapter 6 addresses the policy implications of interactions between distortions.

Calculations in this analysis also do not account for spillovers between sectors. For example, assume that the price of coal is increased to reflect its social cost. All else being equal, the increase would raise the price of electricity and increase demand for alternative fuels. The analysis does not consider substitution both within and between sectors. Similarly, it estimates the effects of power shortages on the productivity of firms, but it does not consider the additional welfare effects on the consumers of goods produced by these firms. The dynamics from removing distortions are also likely to be important. Wisely reinvested revenues from environmental taxes, for example, generate long-run growth benefits. These caveats aside, this report presents a useful first-order approximation of the cost of distortions.

Data limitations preclude the quantification of certain distortions, such as the potential inefficiency of power dispatch in Pakistan and the effects of transmission congestion on power shortages in Bangladesh. Also beyond the scope of this report is a full life-cycle evaluation of the social costs of energy, such as the progression from coal mining through coal transport and waste disposal or the environmental impacts of oil and gas extraction. Exclusion of these impacts will result in underestimation of the cost of distortions but most likely will not affect their ranking.

Data on Utilities, Households, Firms, and More

Most data for this report are compiled from official reports and statistics obtained from government websites or power utilities. Government data are supplemented by household survey data collected by local research institutions and the World Bank, as well as by nighttime light data provided by the U.S. National Oceanic and Atmospheric Administration (NOAA).

UTILITIES

Plant-level productivity is estimated using microlevel data on inputs and outputs of power plants obtained from the annual reports of the Bangladesh Power Development Board (BPDB) for fiscal 2011–14, the annual reports of India's Central Electricity Authority for fiscal 2000–12, and the annual *State of Industry Report* by Pakistan's National Electric Power Regulatory Authority (NEPRA) for fiscal 2006–15. These sources contain plant-level data on installed and derated capacity, vintage, outputs, fuel inputs, and ownership type (publicly owned or independent power producer) for all gas, furnace oil, and diesel units in Bangladesh and Pakistan and for all coal units with a capacity greater than 25 megawatts in India. They also include operating and maintenance costs for plants in Bangladesh and Pakistan. These data are matched with the Platts World Electric Power Plants Database to obtain information on the technology type and turbine makers for power plants.

The production frontier of distribution utilities in India is estimated using data for the distribution sector reported by the Power Finance Corporation, the financial backbone of the country's power sector. These data provide utility-level information on power purchased, own generation, actual sale of power, employee expenditure, capital expenditure, net fixed assets, and organizational structure (such as bundled or unbundled and privatized or corporatized) for 58 distribution utilities for fiscal 2012–16. The data for estimating the efficiency gap in electricity distribution in Pakistan are taken from NEPRA's annual *Distribution Company Performance Evaluation Report*. This report provides for 10 distribution utilities for fiscal 2011–15 utility-level information on total electricity purchased, generated, and sold; peak-load demand; and consumer mix.

In each of these data sets, the same power station or distribution utility can be observed for multiple years. Data sets with this feature, called *panel data sets*, make it possible to control for potential unobserved and time-invariant plant or utility characteristics that could otherwise be omitted from a regression analysis.

HOUSEHOLDS

To estimate the benefits of electrification (or the costs of lack of access to electricity), the analysis relies on data from nationally representative household surveys that include detailed information on households' social and economic characteristics, income, expenditure, education, health, employment, and energy use patterns such as monthly electricity consumption, grid connection status, and appliance ownership. These data sets have two important advantages. First, all of the household surveys used are panel surveys, making it possible to control for unobserved time-invariant household characteristics that could affect both the decision to adopt electricity and the outcomes from gaining access to the grid. Second, in addition to recording a binary variable on electrification status, the surveys asked households to estimate the daily average duration of outages. Using hours of availability as a measure of the reliability of electricity supply, the analysis can therefore estimate the effects of reliability on household welfare.

The household data for Bangladesh are taken from two rounds of a World Bank–sponsored household survey that assessed the effects of grid electrification projects in rural areas of the country. Carried out under the auspices of the Bangladesh Rural Electrification Board in 2005 and 2010, the survey, covers a nationally representative panel sample of 7,352 households in about 1,300 villages.

The household data for India are from the Indian Human Development Survey, carried out jointly by researchers from the University of Maryland and the National Council of Applied Economic Research in New Delhi. The report draws on surveys carried out during 2004–05 and 2011–12, covering a nationally representative sample of more than 20,000 rural households.

The data for Pakistan are from two sources. The first is the Pakistan Social and Living Standards Measurement Survey. The analysis uses surveys from 2007–08 and 2010 because of their panel nature. The sample consists of more than 4,000 rural households. The second source is a household survey carried out by the International Finance Corporation (IFC) under the Lighting Pakistan Program. The survey covered about 6,000 rural households in four provinces. The Pakistan Social and Living Standards Measurement Survey does not collect information on the reliability of power supply, whereas the IFC survey asked about the daily duration of outages in hours.

FIRMS

To estimate the effect of power shortages on firm productivity, the analysis uses micro-level data on firm inputs and outputs from several sources. For Bangladesh, the data are from the Survey of Manufacturing Industries, conducted by the Bangladesh Bureau of Statistics. Although there have been six waves of this survey, only data for fiscal 2012 are used because it is the only year for which firm data can be matched with power shortage data. The fiscal 2012 data cover 8,429 manufacturing establishments.

For India, the firm-level data are taken from the Fourth All-India Census of Micro, Small, and Medium Enterprises, conducted by India's Ministry of Micro, Small, and Medium Enterprises. The census covers about 1.2 million service and manufacturing enterprises, both registered and unregistered, for fiscal 2005–07. About 95 percent are microenterprises, and the median size is two workers. Earlier studies often focused on firms with at least 5 or 10 workers. This data set allows estimation of the effects of power shortages on the smallest firms. Such analysis is important because microenterprises account for a large share of both employment and the gross domestic product in developing countries.

For Pakistan, the firm-level data are from the Census of Manufacturing Industries, conducted by the Pakistan Bureau of Statistics. Four rounds of survey data are available.

However, only fiscal 2011 data are used in order to match them with the power short-age data. The census for 2011 covers 4,500 firms in 23 sectors, mainly in Punjab, the province with the largest economy and the most manufacturing activity.

OTHER DATA

To simulate power plant production profiles, the analysis uses daily dispatch reports from system operators. In Bangladesh, the BPDB provides a detailed daily generation profile for each power plant that includes information on actual generation, available capacity, reasons for shutdown, and duration of plant outages. It also reports a daily demand profile for each substation and daily data on load shedding at the division level. The daily data used cover the period July 24, 2009–December 31, 2015.

The system operator in India, the Northern Regional Load Dispatch Center, provides daily data on power plant production and power shortages. It reports daily information on peak capacity, off-peak capacity, installed capacity, and total generation for each power plant in the country's northern region, as well as daily data on peak shortage, off-peak shortage, and total energy shortage for the entire system. The daily data used by the report cover the period April 1, 2014–March 31, 2015.

Pakistan's system operator, the National Transmission and Dispatch Company, provides daily data on aggregate power generation, aggregate peak-load capacity, aggregate maximum load capacity, minimum load capacity, and nationwide estimated load shedding in megawatts. The data are available for the period February 28, 2014–June 23, 2015.

The analysis also uses many other data sets for various analyses. For example, to detect patterns of power supply disruption at the village level based on nightly varia-tion in brightness, it processed the complete historical archive provided by NOAA of the suborbital Defense Meteorological Satellite Program's Operational Linescan System nighttime imagery captured over South Asia every night since 1993. This archive includes 5 terabytes of data, encompassing some 30,000 visible band images. To esti-mate the effect of cross-subsidies on firm competitiveness, the analysis uses bilateral trade data from the United Nations Comtrade database, as well as energy price data from the International Energy Agency and government reports. Chapters 3–5 provide details on these additional data sets.

References

Clements, Benedict, David Coady, Stefania Fabrizio, Sanjeev Gupta, and Baoping Shang. 2013. *Energy Subsidy Reform: Lessons and Implications.* Washington, DC: International Monetary Fund.

Coady, David P., Ian Parry, Louis Sears, and Baoping Shang. 2017. "How Large Are Global Energy Subsidies?" *World Development* (91): 11–27.

Dahl, Carol A. 2006. "Energy Demand Elasticity Survey: A Primer and Progress Report." In *Energy in a World of Changing Costs and Technologies.* Proceedings of 26th Annual North American USAEE/IAEE Conference, Ypsilanti, MI, September 24–27.

Davis, Lucas. 2014. "The Economic Cost of Global Fuel Subsidies." *American Economic Review: Papers and Proceedings* 104 (5): 581–85.

Parry, Ian, Dirk Hein, Eliza Lis, and Shanjun Li. 2014. *Getting Energy Prices Right: From Principle to Practice.* Washington, DC: International Monetary Fund.

CHAPTER 3

Bangladesh

T he government of Bangladesh has made expanding electricity to the entire population one of its top development priorities. Recent progress toward this goal has been especially impressive. Installed generation capacity has more than tripled since 2009, increasing from 4.9 gigawatts (GW) to 15 GW in 2017. The length of the distribution and transmission lines expanded by roughly 60 percent and 30 percent, respectively, during the same period. Only about 47 percent of Bangladeshis had access to electricity in 2009; by 2017 the share had increased to 80 percent. Thanks to the massive capacity expansion, load shedding has also been drastically reduced, especially since 2014. Power shortages, measured by the share of maximum load shedding to maximum demand, were halved between fiscal 2009 and fiscal 2014, declining from 20.9 percent in fiscal 2009 to 10.1 percent in fiscal 2014, and to 2.2 percent in fiscal 2016 (BPDB 2015, 2016). Because of the increased connection and reduced load shedding, per capita electricity consumption in Bangladesh has almost doubled, rising from 220 kilowatt-hours (kWh) in 2009 to roughly 433 kWh in 2017. The government has also taken initiatives to significantly reduce system losses. Transmission and distribution losses fell from 28.4 percent in fiscal 2001 to 16.9 percent in fiscal 2009 and down to 13.1 percent in fiscal 2016.

In the meantime, Bangladesh is fast becoming a global hotspot for solar home systems as well as solar minigrid and solar pumping energy development, ranking fourth worldwide in the number of workers in renewable energy jobs (IRENA 2017). The government has partnered with nongovernmental organizations and international organizations to promote off-grid options, particularly solar home systems in remote villages where grid electrification is difficult and uneconomic. By 2016 solar panels were generating 150 megawatts (MW) of electricity and serving about 3.5 million households. Installations of solar home systems have averaged more than 50,000 a month since 2009.

Despite this tremendous progress, Bangladesh still ranks low among the world's countries in access to electricity and the reliability of electricity supply. In 2016, more than 38 million of its people were still living without electricity. Its per capita electricity consumption is a third of India's and a tenth of the world average. The 2018 *Global Competitiveness Report* ranks Bangladesh 101st among 137 economies in the reliability of electricity supply (Schwab 2018).

To address these energy supply challenges, Bangladesh intends to rely more heavily on coal. The government's 2010 master plan for the power sector proposed increasing the share of electricity generated from coal to 50 percent by 2030, up from 2 percent in 2015. The latest master plan, from 2016, projected the consumption of coal for power generation would increase from 4.3 million tons in 2015 to 43.1 million tons in 2041. In Bangladesh, the concentrations of smog and fine particulate matter are among the worst in the world. An increasing dependence on coal would exacerbate the situation.

Expanding access to a reliable electricity supply is imperative for economic development in Bangladesh, but increasing the reliance on coal is not the only solution. Many institutional and regulatory distortions have contributed to the electricity shortages, in part by depressing capacity use in gas-based power generation caused by increasing shortages of domestic gas. Addressing these distortions could help boost electricity output while limiting the potential need for coal-based power production.

One notable distortion is the underpricing of natural gas, which artificially stimulates demand and restricts supply. The resulting shortages of natural gas, the main fuel used to generate electricity, were the biggest factor in the low capacity utilization in generation (Ministry of Finance, Bangladesh 2016). With power plants utilized at their optimal level, the existing fleet of generating units would have been sufficient to eliminate the power deficit in fiscal 2016 (Figure 3.1).

FIGURE 3.1 Optimizing the utilization of existing capacity would have been more than enough to end electricity shortages in Bangladesh

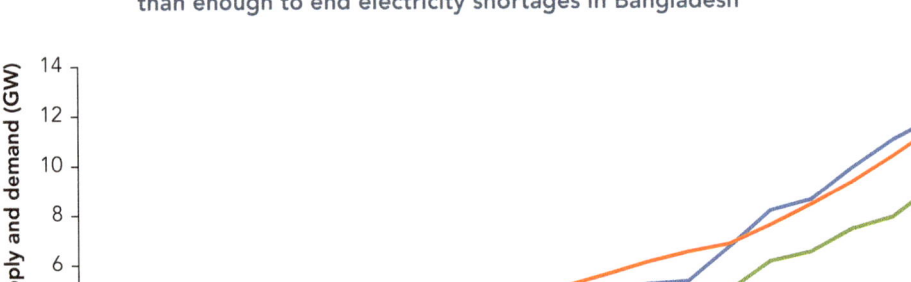

Source: BPDB (2014, 2015, 2016).
Note: FY = fiscal year.

Upstream

Natural gas dominates the energy landscape of Bangladesh. It became increasingly important for energy production after the oil price shock of the early 1970s, when most power plants began diversifying away from oil generation. In 2010, at its peak, natural gas was used to produce 93 percent of the country's electricity (Figure 3.2). That share then fell to 73 percent in fiscal 2016 (BPDB 2016). Electricity generation is the largest source of gas consumption, accounting for 41 percent of the domestic supply in fiscal 2016 (Petrobangla 2016).

Bangladesh's gas supply more than doubled between fiscal 2000 and fiscal 2016, but the existing gas reserves are being rapidly depleted. With compound annual growth of 8.6 percent since the 1980s, annual gas production reached 27.5 billion cubic meters in fiscal 2016, making the country the world's 29th-largest producer of gas. But by December 2015 half of the proven and probable gas reserves, estimated at 27.1 trillion cubic meters, had been depleted.

With declining production from the existing gas fields and a rising demand for electricity, Bangladesh faces increasingly severe gas shortages. The shortfall in domestic production was about 962 million cubic feet a day, or 25 percent, in 2017, up from 621 million cubic feet a day in 2015. Petrobangla, the largest national oil company, projected that, unless significant new production comes online, the demand–supply gap could widen to 1,333.5 million cubic feet a day by 2019 (Petrobangla 2016).

FIGURE 3.2 Natural gas plays a dominant role in electricity generation in Bangladesh

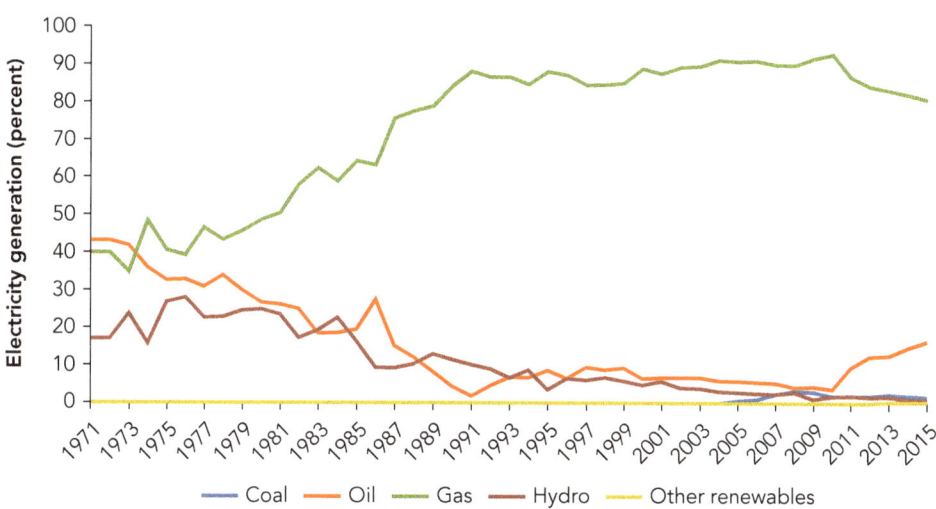

Source: IEA (2017).

Because of the country's heavy reliance on gas for power generation, gas shortages have contributed significantly to electricity shortages. The Bangladesh Power Development Board (BPDB) records the daily operating status of each power station in the country. If a power station is operating below capacity, the power station also reports the contributing factors such as fuel shortages, machine problems, scheduled maintenance or low demand. Calculations based on these daily reports reveal that, on average in 2014, 10 percent of gas-based capacity was stranded because of a gas shortfall (Figure 3.3).

To close the gap left by the domestic gas shortfall, the government resorted to high-cost temporary solutions, including deploying rental power plants running on imported furnace oil and high-speed diesel. Rental plants running on liquid fuel are the most expensive source of power production in both economic and environmental terms. The average cost of these plants is 5–12 times that of gas-based plants, and they are from 30 to 600 percent more polluting, depending on the type of emissions (Figure 3.4).

These plants were set up on an emergency basis, and the goal was to gradually phase them out as permanent plants are installed. But the reliance on rental power plants and liquid fuel has only increased over the years. By 2014 the capacity of rental power plants using furnace oil and diesel had grown to 1,289 MW, or about 12.7 percent of all generation capacity in Bangladesh. The share of oil in the fuel mix for electricity generation increased from 4 percent in 2010 to 16 percent in 2015 (see Figure 3.2).

FIGURE 3.3 Gas shortfalls contributed to significant generation losses in Bangladesh

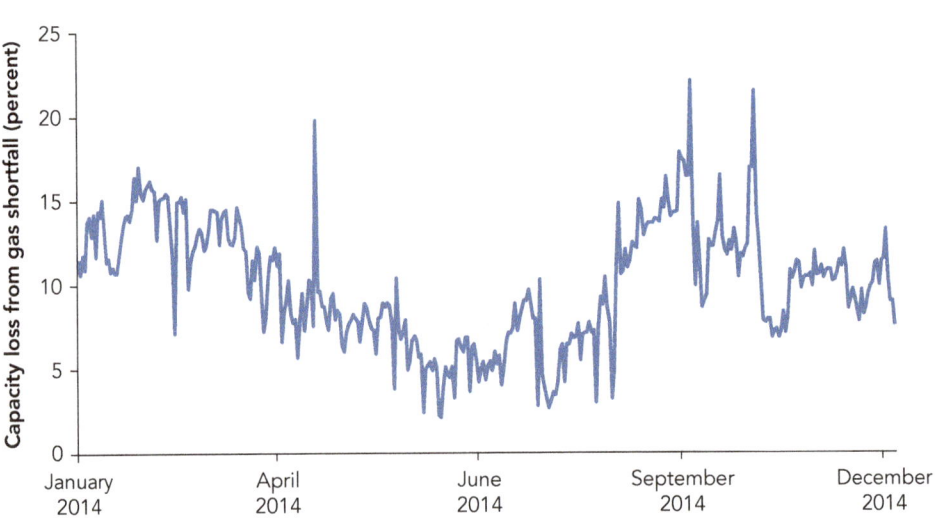

Source: Daily generation reports, Bangladesh Power Development Board, January 1, 2014–December 31, 2014.

FIGURE 3.4 Rental power plants fueled by furnace oil and diesel are much costlier than gas-powered plants in Bangladesh

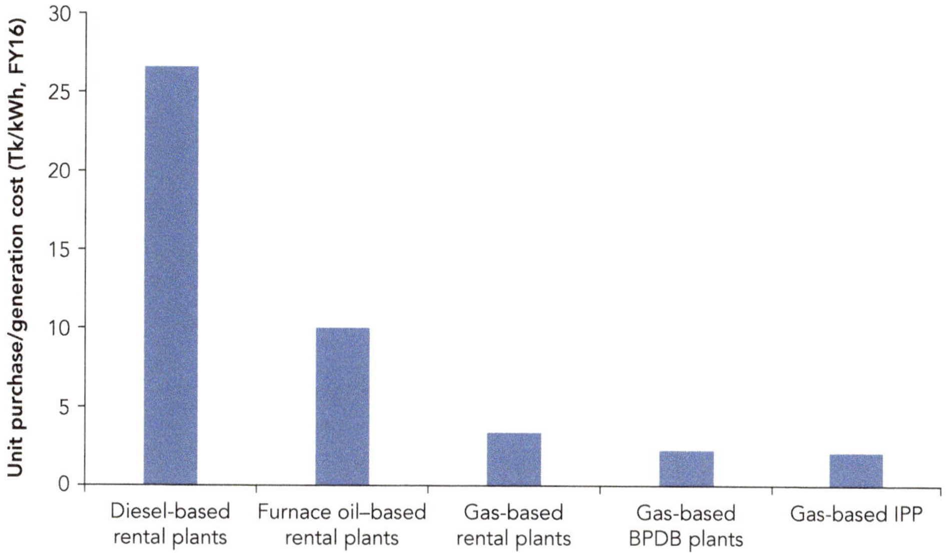

Source: BPDB (2016).
Note: BPDB = Bangladesh Power Development Board; FY = fiscal year; IPP = independent power producer; Tk/kWh = taka per kilowatt-hour.

To deal with domestic gas shortage, the government has also targeted liquefied natural gas (LNG) imports. The first LNG shipment arrived in April 2018. The increasing supply of imported LNG is expected to relieve fuel shortages and cut oil-based power generation.

INSTITUTIONAL: RENTAL POWER PLANTS FAVORED IN GAS ALLOCATION

Petrobangla controls the gas sector in Bangladesh. This vertically integrated national oil company produces about 40 percent of domestic gas, purchasing the other 60 percent from international oil companies. It also controls all gas transmission and distribution.

To cope with acute gas shortages, the government began to ration gas in 2010. It is allocated among power plants, fertilizer plants, industries, and compressed natural gas filling stations in accordance with administrative orders issued from time to time on ad hoc basis rather than market rules. Within the power sector, a committee jointly headed by Petrobangla and BPDB determines how gas is allocated among power stations.

Analysis based on BPDB daily generation reports suggests that the gas allocation scheme has not always been consistent with efficiency goals. The scheme appeared to give priority to gas-based rental power plants, which account for 23 percent of total installed gas capacity in fiscal 2016 and are costlier to operate than plants owned by independent power producers (IPPs) or the government. During 2010–15, gas constraints led to an average capacity loss of 15 percent for the government's own gas-fired plants, whereas the average capacity loss for rental power plants was only 0.77 percent (Figure 3.5). This conclusion holds even when the analysis controls for other confounders such as age, capacity, and common regional and yearly shocks. Even more telling, when plants' fuel efficiency is included as an explanatory variable, along with common regional and yearly shocks, the analysis finds that more efficient plants are more likely to be affected by gas shortfalls. Specifically, a 1 percent increase in a plant's fuel efficiency (defined as the ratio of output power to input gas) is associated with a 1.2 percent increase in the probability of being affected by gas shortages.

Inefficient allocation of gas exacerbates the impact of gas shortages. By contrast, diverting gas from less efficient to more efficient plants could increase electricity output and reduce the unserved energy demand. During hours when there is no power shortage, this increase in production could replace output from uneconomical units such as plants using furnace oil or diesel.

What is the opportunity cost of inefficient gas allocation? To quantify this cost, a scenario is considered in which gas is channeled from plants with lower fuel efficiency to plants with higher efficiency but that were either shut down or are operating below

FIGURE 3.5 Rental power plants are the least affected by gas shortages in Bangladesh

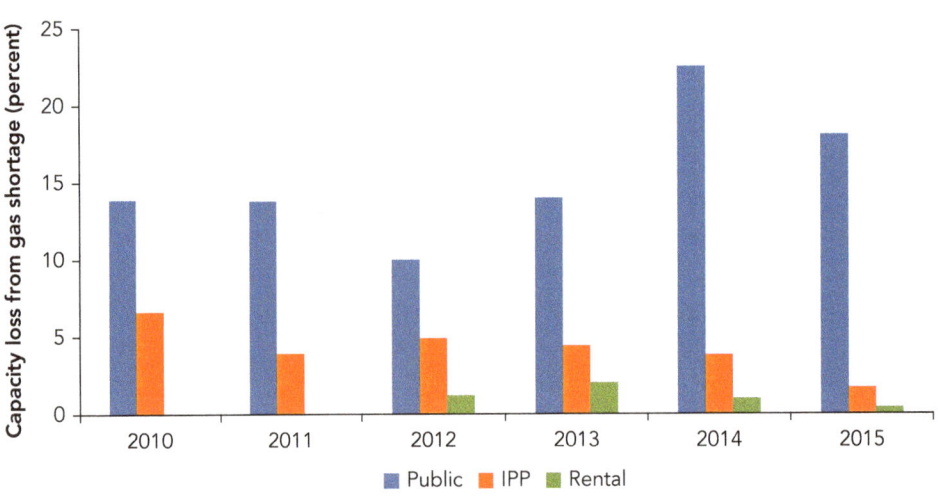

Source: Daily generation reports, Bangladesh Power Development Board, January 1, 2010–December 31, 2015.
Note: IPP = independent power producer.

optimal levels because of gas shortages. The resulting output increases are simulated using half-hourly demand and supply profiles constructed using BPDB daily reports for 2014, the most recent data available at the time of analysis (half-hourly data are a record of electricity demand and supply every half-hour of every day). Assuming a 13 percent loss during transmission and distribution, the remaining output increase is used first to reduce unserved energy demand and then to replace output from more expensive units.

Simulation results show that prioritizing more efficient plants in gas allocation would reduce idled gas capacity by 8.1 percent and reduce the electricity shortage by 15.2 percent a year (Figure 3.6). The use of uneconomical liquid fuel–based units would decline by 4.0 percent (Figure 3.7).

The increase in generation efficiency from better allocation of gas would shift the supply curve of electricity to the right, thereby increasing the electricity supply to the benefit of both consumers and producers. However, government spending on electricity subsidies would also increase because of higher consumption. The net deadweight loss from inefficient gas allocation depends on the shape of the demand and supply curves. On the basis of the estimated long-run supply and demand elasticities for electricity (Box 3.1), the analysis projects a new supply curve based on the simulated increase in electricity supply. It then estimates the new equilibrium quantity at the current level of subsidies and the corresponding change in consumer and producer surplus and government expenditures on subsidies. The net welfare loss from departing from this efficient gas allocation is estimated at $130 million a year.

FIGURE 3.6 **Reallocating gas among existing power plants would have increased gas generation capacity in Bangladesh**

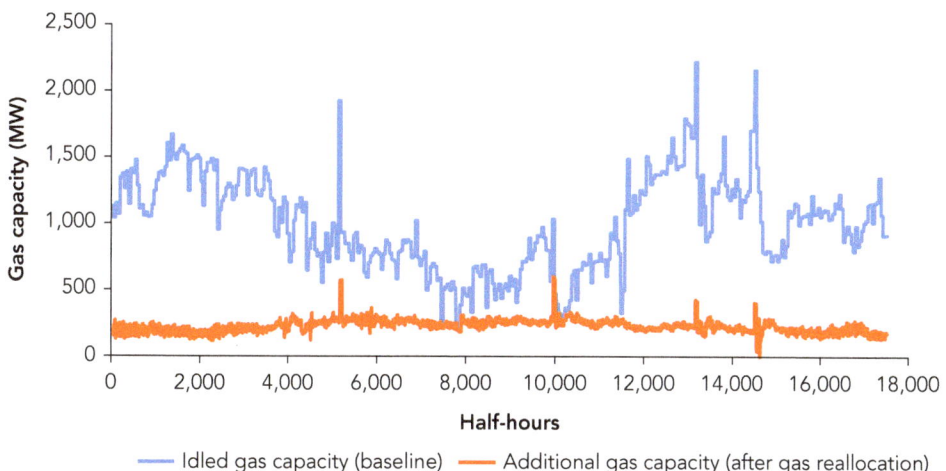

Source: Daily reports, Bangladesh Power Development Board, January 1– December 31, 2014.
Note: MW = megawatt.

FIGURE 3.7 Gas-based electricity output would have increased after gas reallocation in Bangladesh

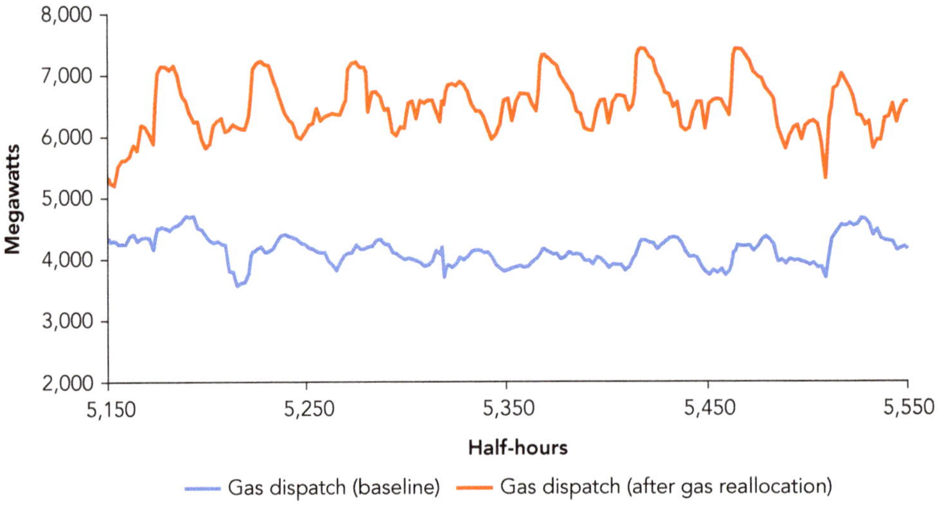

Source: Simulation based on daily reports, Bangladesh Power Development Board, January 1–December 31, 2014.
Note: Figure shows data for part of the year (between the 5,150th and 5,550th half-hour) for illustration. The pattern remains the same for the year as a whole.

BOX 3.1 Estimating the price elasticity of supply and demand for electricity in Bangladesh

This report estimates the price elasticity of electricity supply and demand In Bangladesh. Income and price are considered the main determinants of supply and demand. The estimation is based on annual data for electricity generation, the estimated shortage in the supply of electricity, the real electricity price index, and real per capita gross domestic product (GDP) over the period 1987–2015. The data are from various issues of the annual reports of the BPDB.

To address the potential endogeneity of price, the analysis instruments the contemporaneous electricity price by the first four lags of the electricity price and the first two lags of the price of furnace oil in the estimation of the supply curve. It instruments the contemporaneous electricity price by the first two lags of the electricity price in the estimation of the demand curve. The analysis tests for the existence of long-run cointegration and estimates an error correction model to obtain both short- and long-run elasticities (Appendix A provides details on the methodology).

The estimated long-run supply and demand elasticities are 0.36 and –0.25, respectively. The results indicate that electricity supply and demand in Bangladesh are very inelastic: a 1 percent increase in the price of electricity increases supply by just 0.36 percent and reduces demand by just 0.25 percent. These estimated elasticities are at the lower end of the estimates suggested by the empirical literature on electricity demand in developing countries (Zhang 2015). With higher elasticities, the economic costs of distortions would be even higher.

REGULATORY: UNDERPRICED GAS

The price for natural gas at both the wholesale and retail levels in Bangladesh is regulated and set well below the price of imported LNG or the price for the least-cost replacement fuel. At the wholesale level, Petrobangla purchases gas from both international and national oil companies. The wellhead price for international oil companies is determined by a model production-sharing contract that set the price floor at $100 a ton and the price ceiling at $200 a ton during the latest bidding round for onshore gas in 2012. These prices are equivalent to $20–$40 per barrel of oil, which is substantially below the prevailing international oil price.

Depending on its profit-sharing agreements, Petrobangla can, in addition, receive a large amount of gas at no cost from international oil companies. The share of gas these companies are willing to relinquish for free is a key parameter in their bids for production rights. Still the weighted-average price (based on both purchased and free gas) paid by Petrobangla to the international oil companies was much higher than the purchase price for national oil companies, which operate as monopolies in franchised areas. Because Petrobangla sells gas at the national oil company price to its affiliated distribution companies, the difference between the purchase cost from international oil companies and selling price to distribution companies results in direct gas subsidies.

The retail price of gas is determined by the Bangladesh Energy Regulatory Commission, an independent regulator that oversees the downstream operation of both the gas and electricity sector. This retail price consists of the purchase cost of gas and the cost of gas transmission and distribution. Retail prices are adjusted periodically on the basis of tariff review applications by gas utilities and vary significantly across sectors (Figure 3.8).

The power sector is the largest beneficiary of gas underpricing in Bangladesh. The level of gas underpricing can be illustrated in several ways. One way is to simply compare the price of gas for electricity generation with the cost of supply. At $0.96 per thousand cubic feet in March 2017, this price was 30 percent lower than the weighted-average supply cost of $1.38 per thousand cubic feet reported by distribution companies (both values are in 2015 U.S. dollars)—see Figure 3.8.

Subsidies for gas can also be estimated as the difference between the domestic price and the international benchmark price. Because Bangladesh imports LNG to offset its domestic gas shortfall, a good choice for an international benchmark is the landed import price of LNG at the nearest international hub. Domestic gas is persistently priced below this benchmark (Figure 3.9). Even with the recent price drop in the global gas market, the landed price of LNG to Japan in fiscal 2016 ($7.57 per thousand cubic feet) was still almost seven times the domestic weighted-average price ($1.12) and more than 11 times the price of gas for electricity generation. Because price increases have been too small to offset inflation, the weighted-average retail gas price has been on a declining trend in real terms since fiscal 1992.

FIGURE 3.8 Bangladesh subsidizes the price of gas for the power sector

Source: BOGMC (2015). Estimated cost of supply is based on balance sheets of four major gas distribution companies for fiscal 2014: Titas Gas Transmission and Distribution Company Limited., Bakhrabad Gas Systems Limited, Jalalabad Gas T & D System Limited, and Pashchimanchal Gas Company Limited.

FIGURE 3.9 The weighted-average gas tariff in Bangladesh is significantly below the international price

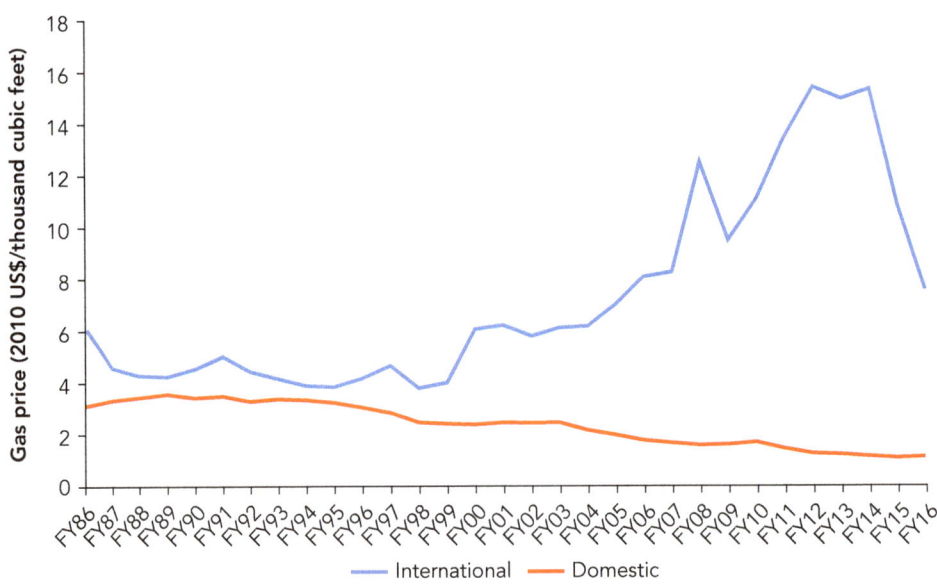

Source: Based on BOGMC (2014, 2015); World Bank Global Economic Monitor Commodities (database); World Bank, World Development Indicators (database).
Note: FY = fiscal year.

Yet another way to gauge the level of gas underpricing is to compare the domestic gas price with the price for the least-cost replacement fuel. For power generation in Bangladesh, this price would be the price of furnace oil and diesel, which is regulated by the Bangladesh Petroleum Corporation, the country's main distributor of oil products. In 2016 the price of furnace oil was roughly 19 times the price of gas for power generation, and the price of diesel was 24 times that price (Figure 3.10).

Beyond imposing costs on the government, gas subsidies artificially stimulate demand and create little incentive for efficient use. Since the 1970s, gas has emerged as the only major fuel for electricity generation in Bangladesh. Without pricing reforms, the government's effort to diversify the fuel mix of electricity generation has not achieved the expected results.

There are also effects on the supply side. Low profitability discourages private investors. International oil companies, though interested in the potential offshore reserves in the Bay of Bengal, have complained about the commercially nonviable gas tariff under production-sharing contracts (Box 3.2).

The underpricing of gas is the biggest source of distortion in the entire value chain of the Bangladesh power sector, leading to large producer losses. Because gas is widely traded on an international market, this analysis considers the international benchmark price to be the market-clearing price. To estimate the deadweight loss created by subsidies, the first step is to estimate the long-run elasticities of supply and demand for gas (Box 3.3). These supply and demand functions are then used to predict what production

FIGURE 3.10 Gas for power generation is much cheaper than the cost of replacement fuels in Bangladesh

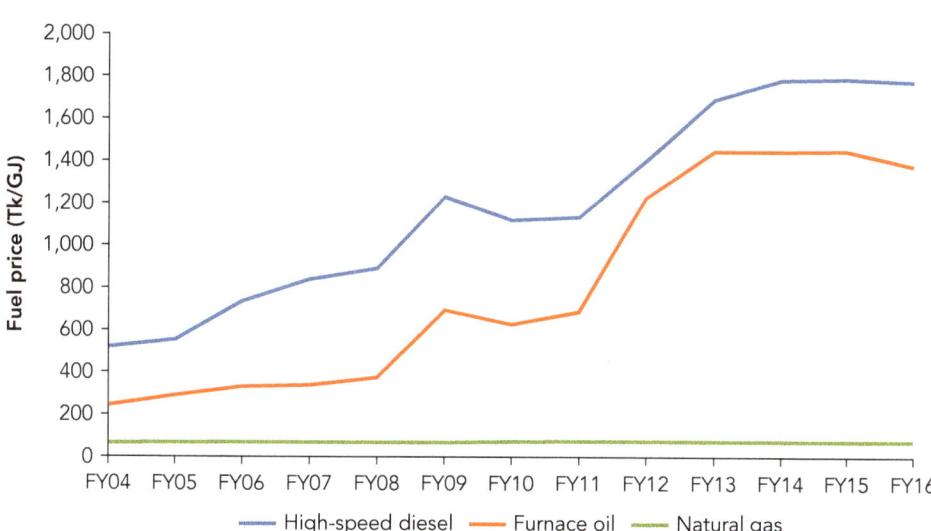

Source: Fuel prices: BPDB (2014, 2015, 2016).
Note: Tk/GJ = taka per gigajoule; FY= fiscal year.

BOX 3.2 Why offshore gas has not been produced in Bangladesh

Bangladesh's offshore gas reserves could be a game changer in resolving the country's gas supply crisis. But an unfavorable price and regulatory framework discouraged offshore exploration and production.

In October 2013, Sangu, the only offshore gas field operating in Bangladesh, was permanently shut down (Reuters 2013). Since then, the country has been relying solely on its onshore gas reserves, which are being rapidly depleted. There is great potential for offshore gas in Bangladesh. Its probable (not proven) offshore gas reserves were estimated at 200 trillion cubic feet (Detsch 2014).

Petrobangla has hosted several bidding rounds aimed at attracting investment to explore the offshore gas potential (Petrobangla 2015). International oil companies explored some offshore blocks under production-sharing contracts (PSC). But, except for the Sangu gas field, none of these efforts turned into actual production. In December 2014, ConocoPhillips pulled out of the operation of two deepwater blocks without drilling exploration wells, despite the company's belief, based on a two-dimensional seismic survey, that the blocks contained 2 trillion cubic feet of gas reserve (Rasel 2014).

Discussion in the media suggests several reasons for the lack of interest among international oil companies. One is the low price in the global gas market, which has made costly offshore exploration less attractive. Another reason is the political uncertainties stemming from maritime disputes with neighboring countries. The third reason may be the rigid terms of the model production-sharing contract designed by Petrobangla. With the contract allowing little or no export, international oil companies have to rely on gas sales to Petrobangla for most of their revenue. But companies perceive the tariff offered by Petrobangla as too low to justify drilling. According to the *Dhaka Tribune*, ConocoPhillips' Bangladesh managing director Thomas J. Earley has argued that the contracted tariff is not commercially viable (Rasel 2017).

Bangladesh is making efforts to gain the interest of international oil companies, especially in offshore exploration and production. The government of Bangladesh has undertaken a review of PSC terms to revise PSC bidding documents and processes. According to recent amendments to the contract, an international oil company would sell almost half of the gas produced to Petrobangla at $6.50 per thousand cubic feet rather than $5.50. Meanwhile, Petrobangla is inviting bids for several more offshore blocks and conducting seismic surveys to provide the market with more information (Petrobangla 2015; Rahman 2016). In 2017, an offshore block was awarded to Posco Daewoo Corporation for gas exploration.

Source: This box was contributed by Weijia Yao, World Bank consultant.

and consumption would be at international market prices. A simulation shows that, at the prevailing international price in fiscal 2016, domestic gas output would more than double whereas demand for gas for power generation would fall to less than one-fifth of current consumption.

On the basis of this simulation, the forgone producer surplus change proportional to gas supply for the power sector is estimated at $5.04 billion a year. This cost is partially offset by the effect on consumers, who benefit from the subsidized gas prices. The net welfare loss is $4.49 billion (2.03 percent of GDP) a year in fiscal 2016.

BOX 3.3 Estimating the price elasticity of supply and demand for gas in
        Bangladesh

In estimating the price elasticity of the domestic gas supply and of the demand for gas by
the power sector, the analysis considers income and price the main determinants of supply
and demand. The estimation is based on annual data for domestic gas production, the
weighted-average retail gas price in real terms (deflated by the consumer price index), the
gas price for the power sector, and real per capita GDP over the period 1980–2010. The
data are obtained from various issues of the annual reports of BPDB and the Bangladesh
Oil, Gas and Mineral Corporation.

To address the potential endogeneity of price, the analysis instruments the
contemporaneous gas price by the first two lags of the gas price and the first four lags of
the real wage index for mining and quarrying in the estimation of the supply curve (using
wage data from various issues of the *Bangladesh Statistical Yearbook*). It instruments
the contemporaneous gas price by its first two lags and the first three lags of the price
of substituting fuel, including furnace oil and high-speed diesel, in the estimation of the
demand curve. The analysis tests for the existence of long-run cointegration and estimates
an error correction model to obtain both short- and long-run elasticities (Appendix A
provides details on the methodology).

The estimated long-run supply and demand elasticities are 0.35 and –0.71, respectively.
Thus in the long run a 1 percent increase in the domestic price of gas increases the supply
of gas by 0.35 percent and reduces the power sector demand for gas by 0.71 percent.
The short-term coefficients on prices are 0.16 for the supply curve and –0.41 for the
demand curve. As expected, short-term elasticities are lower than the long-run estimates
(in absolute terms). Analysis ignoring potential price endogeneity would produce
counterintuitive signs of price elasticities.

Several studies estimate the supply and demand elasticities of gas using rigorous
econometric techniques. Khan (2015) finds that in Pakistan the price elasticity of the power
sector demand for gas is –0.51 in the short run and –0.76 in the long run. Hausman and
Kellogg (2015) find that in the United States the long-run price elasticity of the gas supply
is 0.81 and the long-run price elasticity of the power sector demand for gas is –0.47. The
estimated price elasticities in this report are in the range of the elasticities suggested in
the literature.

The government could have used this large amount of forgone revenue to invest in
infrastructure, education, health care, and other public services to promote long-term
economic growth. The regulatory cost of gas underpricing would be even higher if its
adverse effect on long-term growth were considered.

In addition to the price of gas, the government also regulates the prices of furnace oil and diesel. At times, these administered prices have been set below the
import cost, resulting in large losses (under-recoveries) for the Bangladesh Petroleum
Corporation, which incurred losses every fiscal year between 1990 and 2013. But,
because the recent fall in international oil prices was not passed through to consumers, the company was able to generate surpluses from the sale of petroleum products
from fiscal 2014 to fiscal 2017. During this period, the domestic price of petroleum

products was higher than the international spot market price even after including transport, distribution, retailing, and the negative environmental externalities of fuel consumption (World Bank 2016). However, because the international oil price rose in 2017, the Bangladesh Petroleum Corporation is once again facing losses. Just as there are deadweight losses associated with pricing below cost, so there are deadweight losses associated with pricing above cost. Data limitations preclude estimation of the welfare losses from price distortions for petroleum products, however.

SOCIAL: EMISSIONS FROM THE USE OF GAS AND OIL

Natural gas has been widely promoted as a cleaner fuel, but its combustion still produces greenhouse gas emissions and contributes to urban smog. In 2015 natural gas combustion for electricity generation accounted for 40 percent of greenhouse gas emissions in Bangladesh. Burning gas also produces nitrogen oxides, which are precursors to ozone, the main component of smog. Over the past 25 years, Bangladesh has experienced some of the world's largest increases in seasonal average population-weighted concentrations of ozone. A toxic compound and dangerous irritant, ground-level ozone can cause respiratory ailments such as chronic obstructive pulmonary disease. In 2015 ambient ozone contributed to 7,900 deaths in the country, an increase of more than 200 percent since 1990 (Health Effects Institute 2017).

When pricing of natural gas fails to account for these external costs, it results in consuming too much gas to generate power. To estimate the associated social cost, the analysis calculates the change in welfare that would occur when the externalities of gas consumption are internalized, achieved by introducing an environmental tax. Parry and others (2014) estimate that gas combustion in Bangladesh led to climate change damages of $2.41 per gigajoule (GJ) and health damages of $0.21 per GJ. Imposing an environmental tax equal to the sum of these marginal damages, with an estimated demand elasticity of −0.71, would reduce annual gas consumption for power generation by 256.68 million GJ. The external cost of this excess consumption, at $2.62 per GJ, would amount to $0.67 billion annually. Higher gas prices would also lead to a reduction in the consumer and producer surplus, partially offset by the increased tax revenue. The net change in welfare (or net social cost)—approached as the sum of avoided environmental and health damages, increased revenue from environmental taxation, and forgone consumer and producer surplus—would be $345 million a year.

Upstream social distortion also results from excessive use of diesel and furnace oil associated with the inefficient allocation of gas. Burning liquid fuel is more polluting than burning gas. More efficient allocation of gas would offset the need for oil-based power generation and thus reduce emissions.

How big are the potential benefits? Using data on fuel consumption, heat efficiency, and the operational status of all generating units in Bangladesh, the analysis performs a simulation in which gas is channeled from plants with lower efficiency to more efficient plants that were affected by gas shortages. This reallocation would reduce annual emissions of carbon dioxide from diesel- and furnace oil–based plants by 4 percent, or 0.25 million tons. On the basis of the shadow price of carbon dioxide emissions of $40 per ton, the avoided external costs would amount to $10 million a year.

Core

Since its beginning, the core segment of the power sector has been controlled and dominated by the BPDB, which was established in 1972 as a vertically integrated utility responsible for the generation, transmission, and distribution of electricity throughout the country. The Rural Electrification Board was created in 1977 to expand electricity supply in rural areas. The Dhaka Electricity Supply Authority, established in 1991, is responsible for power distribution in and around Dhaka.

By 1994 the country's public power sector was in extremely poor condition. Average distribution losses had reached 37.2 percent, and collection-to-billing ratios were low, at 62.4 percent for BPDB and 54.6 percent for the Dhaka Electricity Supply Authority. Only the Rural Electrification Board was faring better, with distribution losses of roughly 15 percent and a collection ratio close to 100 percent. There were also acute power shortages. Average load shedding amounted to close to a fifth of total demand during fiscal 1991 and 1995.

In response to this crisis, the government launched a major power sector reform in 1994. Key elements included unbundling and corporatizing the operating units of BPDB, corporatizing the Dhaka Electricity Supply Authority, boosting private sector participation in generation, and setting up an independent regulatory body, the Bangladesh Energy Regulatory Commission.

The reform substantially reduced distribution losses, shaving them from 26 percent in fiscal 2000 to 10 percent in fiscal 2017. Collection ratios exceeded 95 percent in fiscal 2016 for all distribution companies except the Dhaka Power Distribution Company, which had a ratio of 64 percent, and as of fiscal 1996 net revenue collection increased more than tenfold.

But electricity blackouts continued to worsen until fiscal 2013. Despite a tripling of installed generation capacity in the public and private sectors between fiscal 2000 and fiscal 2013, the peak demand shortage increased from 200 MW to 1,048 MW in fiscal 2013 over the same period. Power shortages have, however, been largely reduced over the past few years because of the introduction of additional generation capacity (including from rental power plants) and the fall in the global price of oil, which has made furnace oil–based units less expensive to run. Maximum load shedding stood at 250 MW in fiscal 2016.

Besides gas supply constraints, several other factors also contributed to the electricity shortage, including inefficient generation, non-merit-based electricity dispatch, and the underpricing of electricity.

INSTITUTIONAL: INEFFICIENT GOVERNMENT-OWNED POWER PLANTS

Despite growing participation by the private sector, publicly owned utilities still play a major role in electricity generation in Bangladesh. In fiscal 2016, the public sector accounted for more than 50 percent of installed generation capacity and 43 percent of electricity production (Figure 3.11). Public power plants are of two types: those owned and directly controlled by BPDB and those owned by BPDB subsidiaries, which are corporatized and thought to have more management autonomy.

FIGURE 3.11 The public sector accounts for a large share of electricity generation in Bangladesh

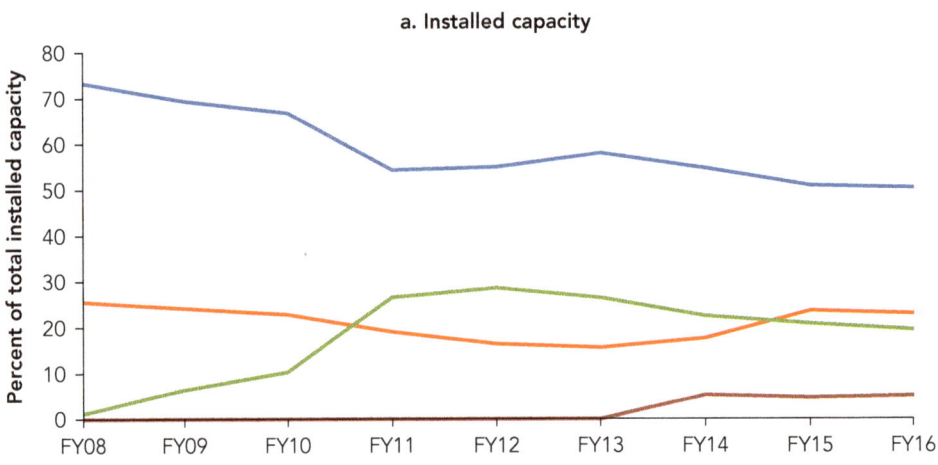

a. Installed capacity

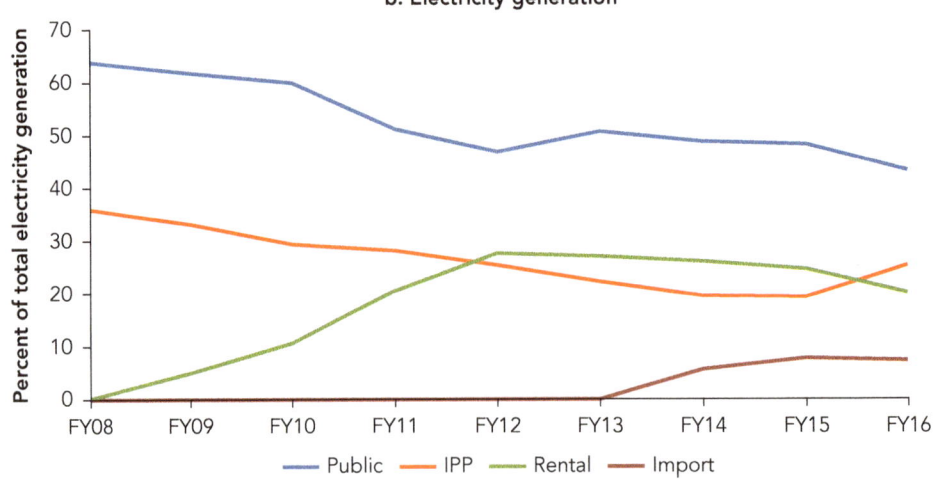

b. Electricity generation

Source: BPDB (2016).
Note: FY = fiscal year; IPP = independent power producer.

Private sector plants can also be roughly categorized into two types: rental plants (including rentals and quick rentals) and plants owned by IPPs. Rental plants have played an increasingly important role since first being synchronized to the grid in fiscal 2008; their share of both capacity and electricity production had grown to 20 percent by fiscal 2016. By contrast, although the installed capacity of IPPs more than doubled between fiscal 2008 and fiscal 2016, their share of capacity stayed roughly the same, at about 23–26 percent, and their share of output fell sharply, from 36 percent to 25 percent.

In fiscal 2016, 56 percent of rental capacity was based on gas, about 40 percent on furnace oil, and 4 percent on high-speed diesel. But rental plants also represent the largest share of liquid fuel–based capacity, accounting for about 40 percent of diesel- and furnace oil–based capacity overall (Figure 3.12).

For all plants, regardless of ownership type, the current approach to tariff setting provides little incentive to increase efficiency. Power plants are allowed to pass fuel costs to consumers under rate-of-return regulation. In addition, plants that have power purchase agreements with the BPDB are guaranteed a fixed monthly payment (the so-called capacity payment), even when they generate no electricity. Because inefficient behavior is unlikely to be penalized, incentives to improve efficiency are weak, especially at public sector utilities, which lack a strong profit motive.

Do the institutional and contractual arrangements facing different types of plants affect their efficiency? The analysis explores this question using output and input data for all power-generating units in Bangladesh from fiscal 2011 to fiscal 2014. The data reveal a large dispersion in plants' fuel efficiency as measured by the capacity factor

FIGURE 3.12 Rental plants represent the largest share of liquid fuel–based capacity in Bangladesh

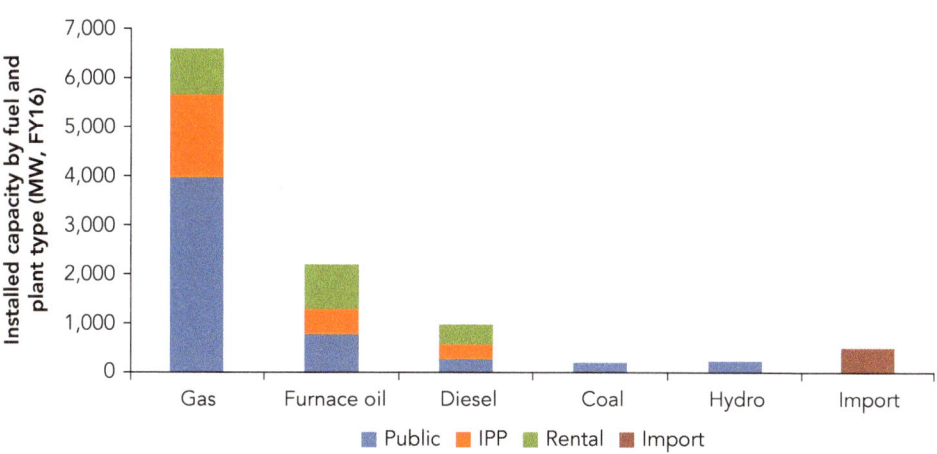

Source: BPDB (2016).
Note: FY = fiscal year; IPP = independent power producer; MW = megawatt.

(the ratio of actual output to maximum feasible output) and the ratio of electricity output to heat input (Figure 3.13). Unsurprisingly, IPPs have the highest capacity factor and are the most efficient at converting fuel into electricity. Private rental units have the next highest efficiency, followed by corporatized units and BPDB units.

FIGURE 3.13 Public power plants are less efficient than private plants in Bangladesh

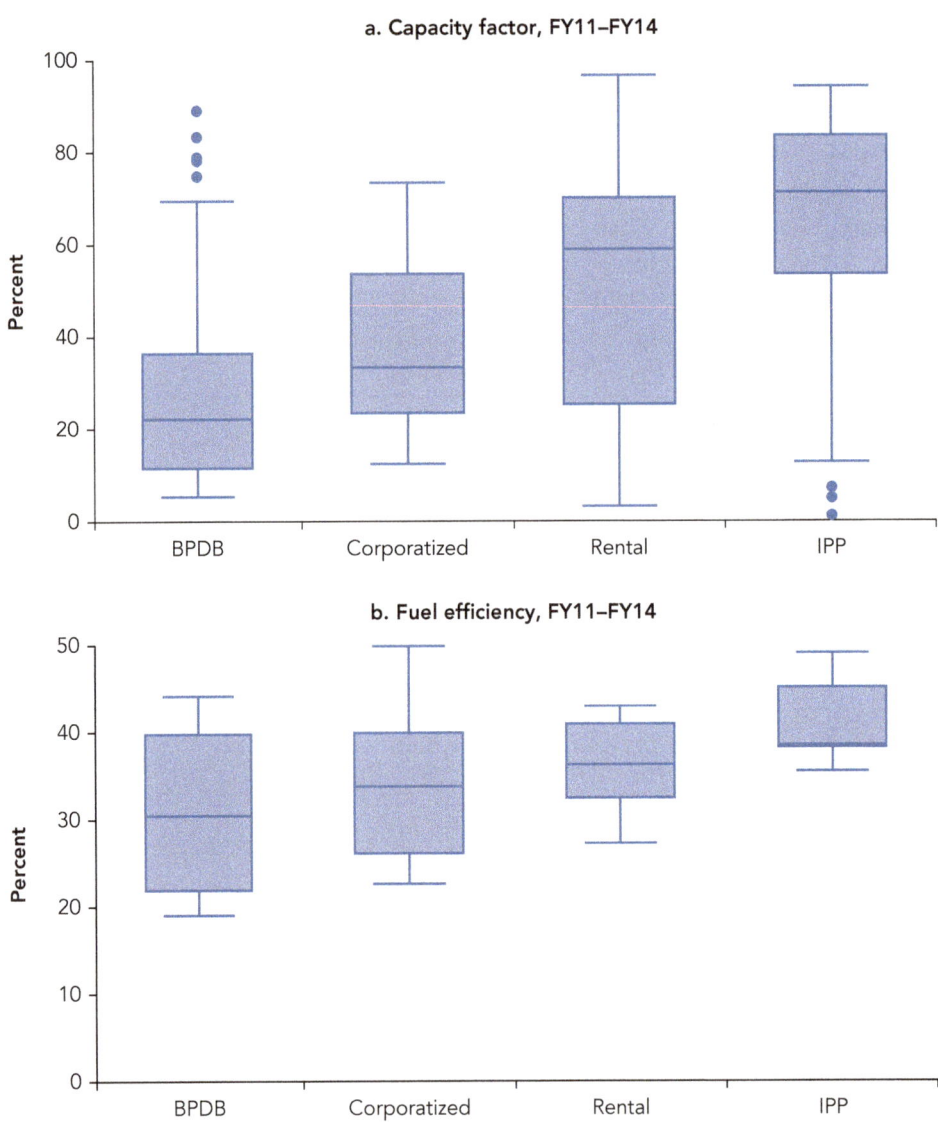

Source: Based on plant-level data, BPDB (2011–14).
Note: Fuel efficiency is the ratio of electricity output to heat input. The figures compare the median values of plants' capacity factor and fuel efficiency without controlling for other confounding factors. BPDB = Bangladesh Power Development Board; FY = fiscal year; IPP = independent power producer.

Many factors could explain the differences in efficiency observed among different types of plants. For example, IPPs are newer and larger and are more likely to be combined-cycle units. To separate the effects of institutional arrangements and physical attributes, the analysis estimates an input demand function for fuel (measured in gigajoules) while controlling for plants' total output, vintage, size, fuel type, geographic location, technology (gas turbine, internal combustion engine, steam turbine, or combined cycle, with combined cycle being the base case), and common yearly shocks, as well as for their institutional type (BPDB, corporatized, rental, or IPP, with IPP being the base case).

The results show that, even after controlling for these potential confounding factors, BPDB, corporatized, and rental power plants are still less efficient than IPPs (Figure 3.14). All else being equal, on average corporatized units use 37 percent more energy than IPPs to produce the same amount of electricity, BPDB units use 25 percent more, and rental units use 17 percent more. These numbers reflect the average efficiency gap across all types of fuel. Restricting the sample to gas plants reveals an even greater difference in fuel efficiency between BPDB and IPP plants—29 percent.

FIGURE 3.14 Public power plants in Bangladesh are less efficient than private plants even after controlling for their characteristics

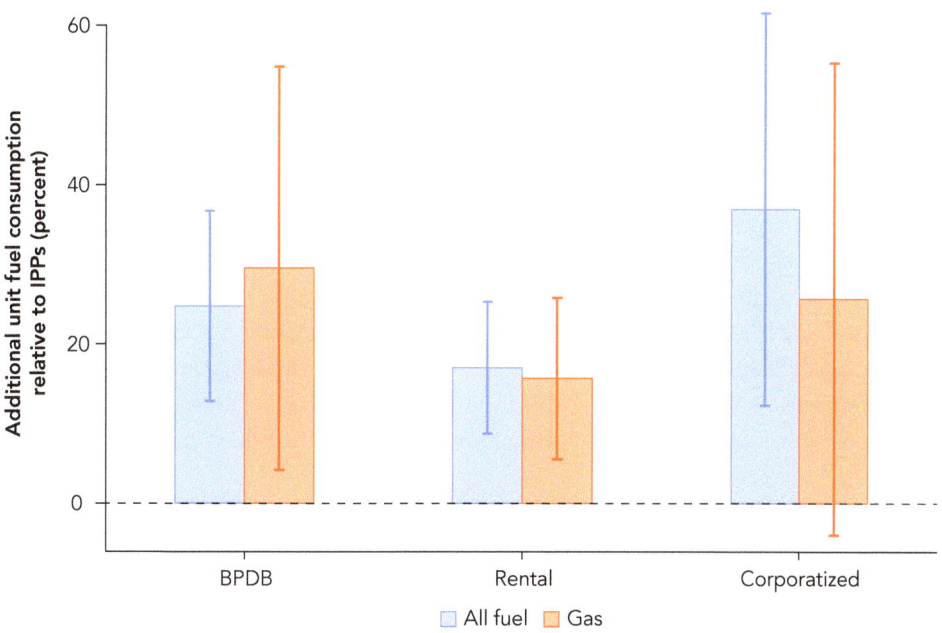

Source: Based on plant-level data, BPDB (2011–14).
Note: Fuel efficiency is measured as fuel input (in gigajoules) per unit of electricity produced. The figure shows the relative difference in fuel efficiency between IPPs and other types of plants after controlling for plants' physical and technological characteristics. Bars denote point estimates and lines denote 95 percent confidence interval. BPDB = Bangladesh Power Development Board; IPP = independent power producer.

Notably, fuel efficiency could also be affected by how often a unit is shut down and ramped up and by how long it is operated below capacity. Shutdowns and underutilization may reflect inefficiency (for example, poor operating and maintenance practices may lead to machine malfunctions and forced shutdowns), but they can also result from a disadvantage in getting access to fuel or from dispatch that does not always follow merit order (in which the most efficient plants are called on first). One indicator used as a proxy for how often a plant is called on to provide electricity is the capacity factor. But, because the capacity factor is determined at least in part by a plant's fuel efficiency, it cannot be directly included as an explanatory variable for fuel efficiency. To address this simultaneity concern, the analysis follows Fabrizio, Rose, and Wolfram (2007) in using regionwide electricity demand as a source of exogenous variation in the utilization rate of peak-load gas units. Regardless of its institutional type, a peak-load plant is more likely to be running when there is a regionwide demand surge, but regionwide demand is unlikely to be correlated with an individual plant's fuel efficiency.

The basic conclusion remains the same after the sample is restricted to peak-load gas units and regionwide demand is used as a surrogate variable for plants' capacity factor (Figure 3.15). After the analysis controls for plants' dispatch, the efficiency gap between IPP and rental units becomes both larger and more statistically significant. Everything else being equal, a rental unit would use 21 percent more gas than an IPP unit to produce the same amount of electricity. Consistent with results elsewhere, this finding may suggest that, relative to IPPs, rental power plants are favored in access to fuel and possibly in dispatch. The efficiency gap for BPDB and corporatized units is also larger, though less precisely estimated because of a smaller sample size.

One interpretation of these results is that differences in fuel efficiency may be driven by differences in the fuel prices charged to different types of plants. Plants facing lower fuel prices may have used fuel more intensively (relative to other types of inputs). For example, BPDB guarantees rental power plants access to fuel at prices lower than those the Bangladesh Petroleum Corporation charges other consumers. For diesel, the difference is estimated at 42 takas (Tk) per liter, paid by BPDB (World Bank 2015a). To assess this interpretation, the analysis estimates the correlation between institutional type and total factor productivity/technical inefficiency, which accounts for variation in output while holding all observable inputs fixed. Producers with higher total factor productivity will generate more electricity regardless of differences in factor prices for fuel, labor, and capital.

The stochastic frontier approach is used to estimate a production function, and technical inefficiency is defined as one of the error terms of the stochastic production frontier. The strength of the stochastic frontier approach is that it considers stochastic noise in the data and estimates both idiosyncratic productivity shocks and technical inefficiency.

With all observable inputs considered, IPPs continue to lead in efficiency. Their technical efficiency scores are significantly higher on average than those of the other three types of plants (Figure 3.16). Perhaps surprisingly, the technical efficiency

FIGURE 3.15 BPDB and rental gas plants in Bangladesh are less efficient than IPPs, even after controlling for plants' dispatch

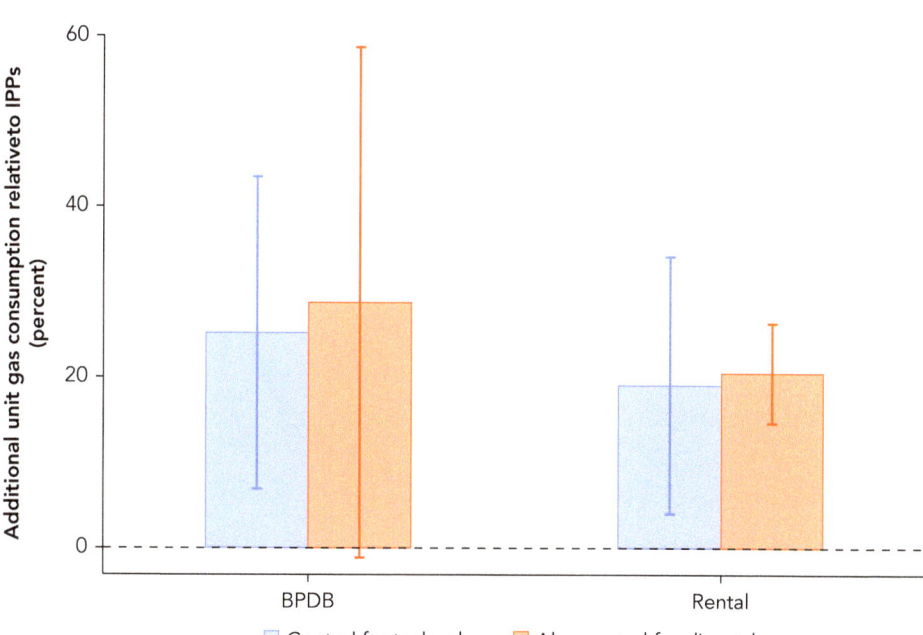

Control for technology Also control for dispatch

Source: Estimation based on plant-level data, BPDB (2011–14).
Note: Fuel efficiency is fuel input (in gigajoules) per unit of electricity produced. The figure shows the relative difference in fuel efficiency across plant types after controlling for plants' physical and technological characteristics and dispatch. Orange bars show estimated coefficients from regression analysis controlling for dispatch through the control of statewide electricity output. Blue bars show estimated coefficients from regression analysis controlling for plants' physical and technological characteristics but not for dispatch. Lines denote 95 percent confidence interval. Coefficients for corporatized units are estimated but not reported because of the small sample size. BPDB = Bangladesh Power Development Board; IPP = independent power producer.

scores of the BPDB and corporatized units are statistically indistinguishable even at the 10 percent level. The same is true for the scores of the BPDB and rental units. Overall, the average score for all plants is estimated at 0.77, indicating that the actual output of power plants in Bangladesh is on average about 77 percent of their maximum feasible output.

Because the analysis controls for the exogenous physical, technical, and operational characteristics of power plants, the effects of institutional settings on generation efficiency are most likely caused by other innate differences between plants. Some of these differences may be explained by the type of power purchase agreements signed by private plants. Anecdotal evidence suggests that some private plants are allowed to be dispatched only at the optimal load factor to ensure maximum fuel efficiency. But, because power purchase agreements are mostly confidential, the extent of such preferential treatment is unknown.

FIGURE 3.16 IPP plants are the most efficient power plants in Bangladesh

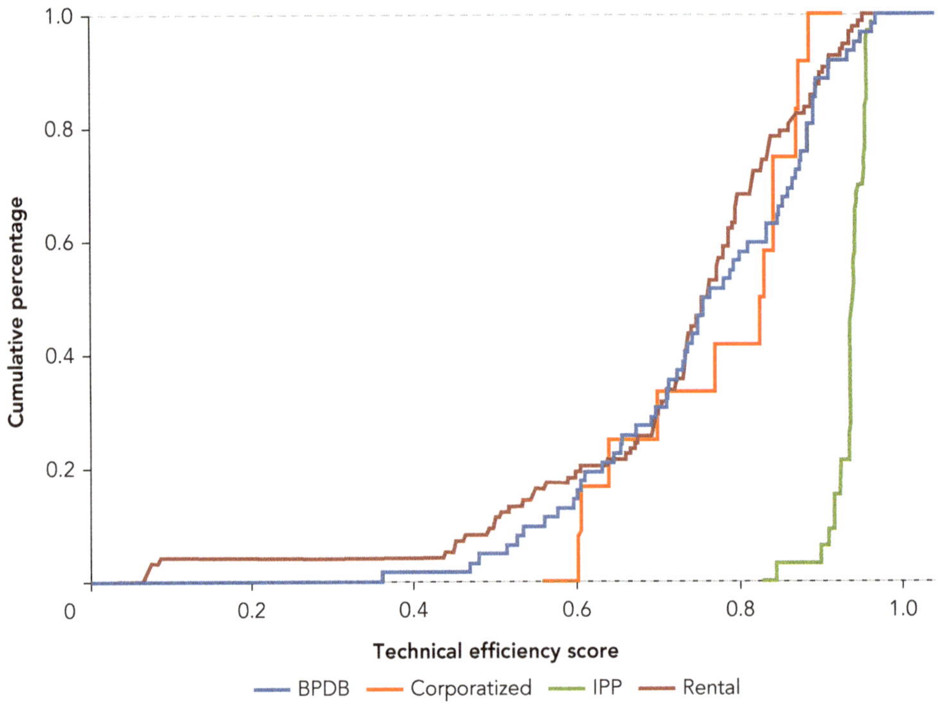

Source: Estimation based on plant-level data, BPDB (2011–14).
Note: The technical efficiency score ranges between 0 and 1. It measures the ratio of actual output to maximum feasible output. BPDB = Bangladesh Power Development Board; IPP = independent power producer.

To the extent that the efficiency gap between BPDB/rental power plants and IPPs is also related to differences in the quality of managerial practices across ownership types, there could be large potential for institutional reform to improve generation efficiency. Improving generation efficiency would be a cost-effective way to reduce power shortages even under gas supply constraints. A simulation is conducted to estimate how much more electricity could be produced if gas-based BPDB and rental units were to match the fuel efficiency of IPPs (after controlling for differences in physical and technological attributes). BPDB plants would have to increase their fuel efficiency by 29 percent and rental plants by 21 percent, all else being equal. The simulation constructs half-hourly dispatch of power plants while taking into account existing operational constraints, such as whether a plant is unavailable because of scheduled maintenance or mechanical failures. The simulation assumes that each plant uses its own fuel savings from efficiency improvement up to its maximum capacity. The additional production, subject to a 13 percent network loss, is used first to reduce unserved energy demand and then to replace output from more costly units, such as those using diesel and furnace oil.

The simulation results show that efficiency improvement alone would have increased gas-based power generation by 4.5 percent and reduced power shortages by 49 percent in 2014 (Figure 3.17). Meanwhile, the increase in output from gas units would allow oil-based units to be dispatched less often, reducing oil consumption by 10 percent. This increase in fuel efficiency would shift the supply curve to the right. Similar to the earlier analysis of the inefficiency of gas allocation, the analysis predicts new supply and new equilibrium consumption based on the estimated long-run supply and demand elasticities for electricity (see Box 3.1). The net total welfare cost of inefficient electricity generation is estimated at $350 million a year.

If Bangladesh were to promote greater competition in electricity generation, the potential gains from efficiency improvement would be even higher. Studies of the restructuring of the U.S. electricity market show that even IPPs became more efficient after the industry moved from rate-of-return regulation to a market-based tariff regime (Fabrizio, Rose and Wolfram 2007; Zhang 2007).

Since the launch of power sector reform in 1994, Bangladesh has turned around the performance of electricity transmission and distribution. The average loss was reduced from 32 percent in 2000 to 13 percent in 2016 (Figure 3.18).

FIGURE 3.17 Improving generation efficiency reduces power shortages and the output of oil-based plants in Bangladesh

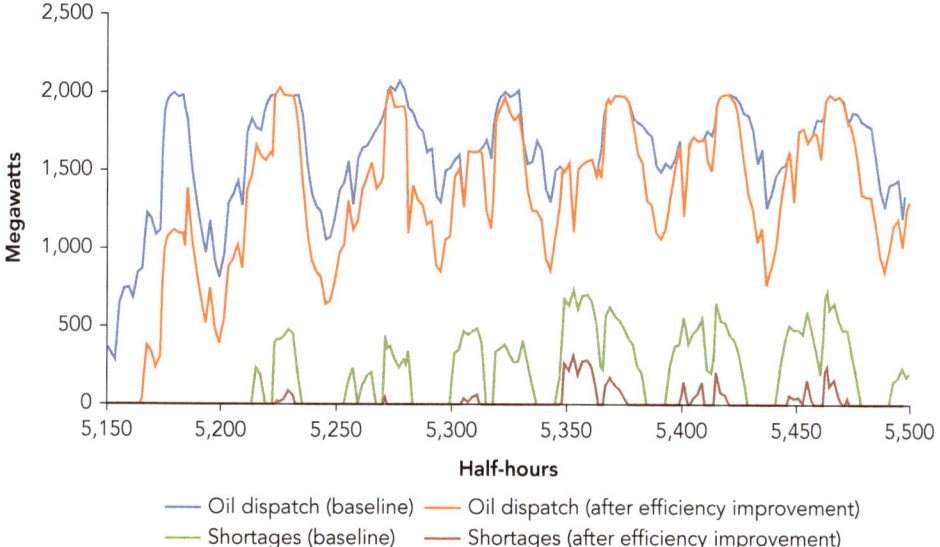

Source: Simulation based on daily reports, Bangladesh Power Development Board, January 1–December 31, 2014.
Note: Figure shows data for part of the year 2014 (between the 5,150th and the 5,550th half-hour). The pattern remains the same for the year as a whole.

FIGURE 3.18 Electricity transmission and distribution losses declined between 2000 and 2016 in Bangladesh

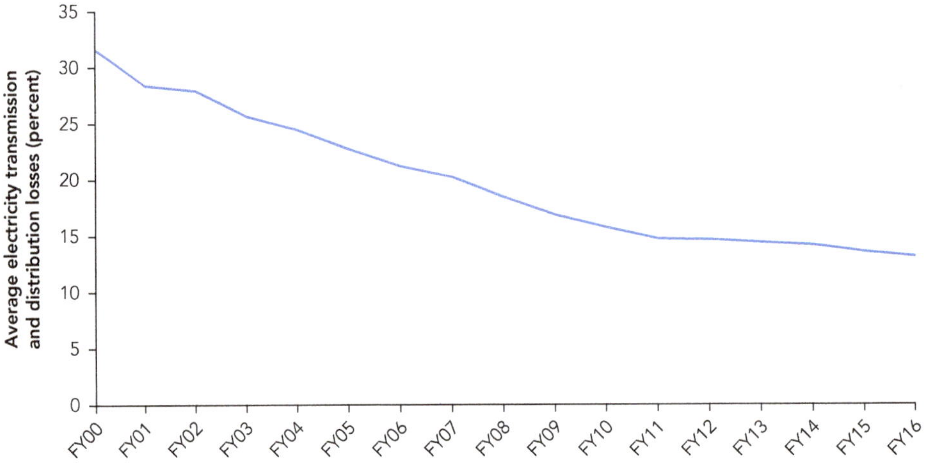

Source: BPDB (2016).
Note: FY = fiscal year.

This impressive achievement notwithstanding, more could be done to bring sector performance in line with the international average of 10 percent loss. In the Mymenshing Division of Bangladesh, the distribution loss alone has hovered around 15 percent in recent years. Reducing the average transmission and distribution loss to 10 percent would shift the supply curve to the right, increasing both consumer and producer surplus. Excluding increases in subsidy spending, the net deadweight loss from less efficient electricity transmission and distribution is estimated at $60 million a year.

INSTITUTIONAL: DISPATCH NOT BASED ON MERIT ORDER

The Bangladesh power sector follows a single-buyer model: BPDB acts as the central agent, purchasing power from public and private generators and selling it through bulk sales to all distribution entities (including its own distribution companies). A single-buyer model makes dispatch less transparent and more vulnerable to interventions that could result in out-of-merit dispatch in which less efficient plants are dispatched earlier than more efficient plants.

A recent World Bank study using actual hourly dispatch data from the National Load Dispatch Center of Bangladesh for all of 2014 finds that merit order was not used in dispatching generators (World Bank 2015b). Liquid fuel–based rental power plants were often brought into production before other lower-cost generators. Using a dispatch optimization model, the study simulates real-time system operation hour by hour (considering both the marginal cost of production and distance to load centers to minimize transmission losses) to produce the least-cost dispatch. It shows that actual dispatch deviates substantially from optimal dispatch. On January 1, 2014, for example, oil-based units with costs above Tk 5 and even Tk 20 per kWh were dispatched even when less costly options (gas, coal, and import capacity) were available (Figure 3.19).

Improving dispatch to ensure that the lowest-cost generators are used as much as possible before more expensive generators are brought online could reduce the cost of power generation. On the basis of 2014 data, the same World Bank study estimates that bringing dispatch to the optimal level would reduce production costs by $1.65 billion a

FIGURE 3.19 More expensive oil-based units were dispatched in Bangladesh even though less costly capacity was available

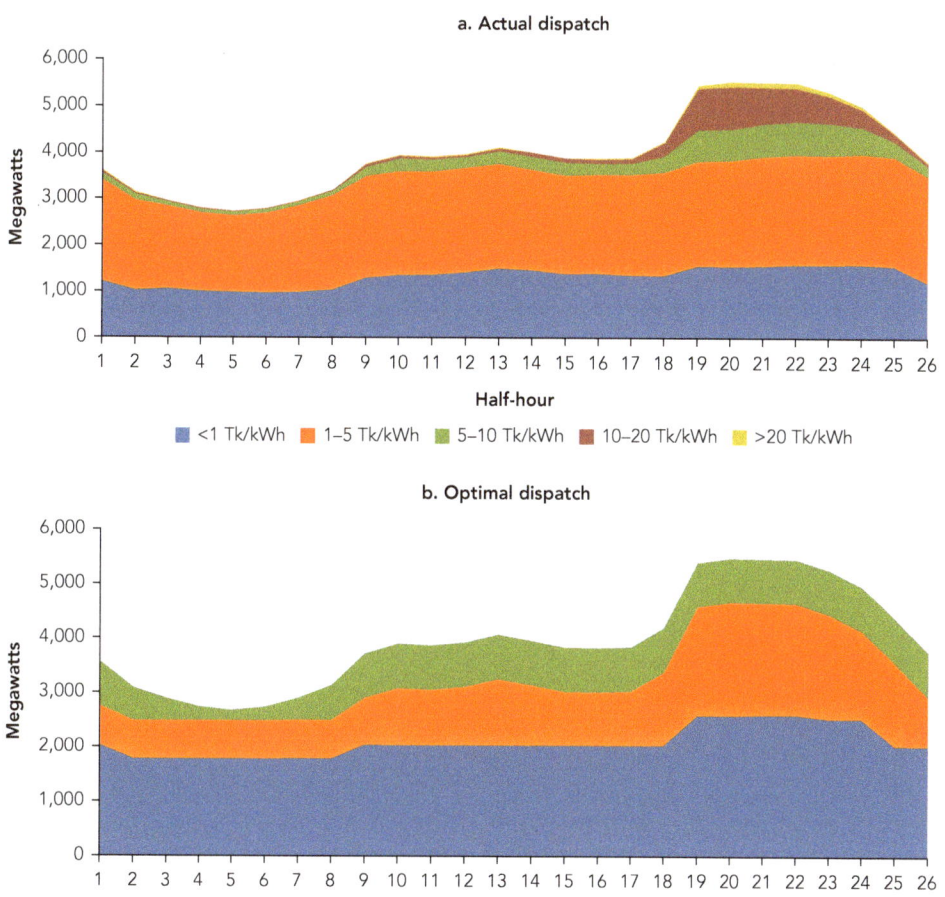

Source: World Bank (2015b).
Note: There are 26 rather than 24 time-steps because the evening hours from 6 to 8 are half-hourly time-steps. kWh = kilowatt-hour; Tk = Bangladesh taka.

year with no increase in gas consumption. Improving dispatch efficiency may require some training and investment—to replace the current spreadsheet-based manual dispatch with an automated dispatch model, for example, and to increase investment in transmission capacity to eliminate congestions that have prevented evacuation of lower-cost plants in certain locations. But the costs would be small compared with the potential savings.

REGULATORY: UNDERPRICED ELECTRICITY

Bangladesh not only subsidizes the cost of electricity production through underpriced gas but also directly subsidizes electricity tariffs at the wholesale and retail levels. At the wholesale level, BPDB has sold electricity to distribution companies at a price below its purchase cost (Figure 3.20). The growth in the share of liquid fuel–based power generation has pushed up both the average bulk supply cost of electricity and subsidies, which surged from Tk 45 billion in fiscal 2011 to Tk 60 billion in fiscal 2014 and 2015. These subsidies were financed through cash loans at subsidized interest rates and through direct budgetary support.

At the retail level, electricity rates vary by consumer category, volume of consumption, and time of use (peak or off-peak) and are subsidized for households and farmers. These subsidies are financed in part through the higher rates for industrial and commercial users (cross-subsidies)—see Figure 3.21. However, the weighted-average retail tariff is still below the already subsidized wholesale price, resulting in losses for distribution companies. The government makes up this shortfall through direct budgetary transfers, which reached Tk 13.3 billion (in fiscal 2010 prices) in fiscal 2012.

FIGURE 3.20 Electricity is sold to distribution utilities below its purchase cost in Bangladesh

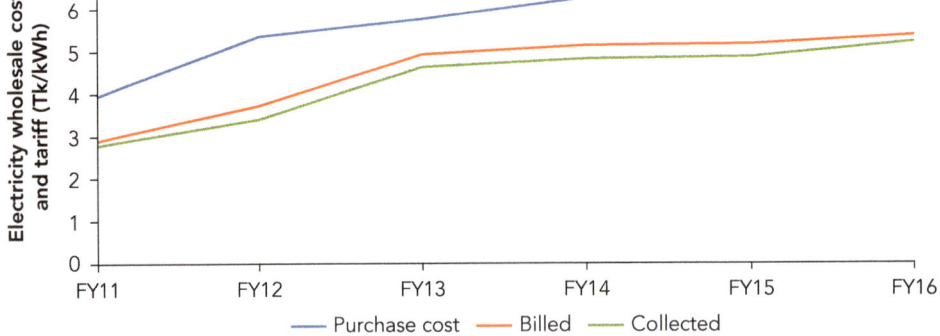

Source: Calculations based on BPDB (2015, 2016).
Note: FY = fiscal year; kWh = kilowatt-hour; Tk = Bangladesh taka.

FIGURE 3.21 Industrial and commercial users subsidize residential and agricultural users in Bangladesh

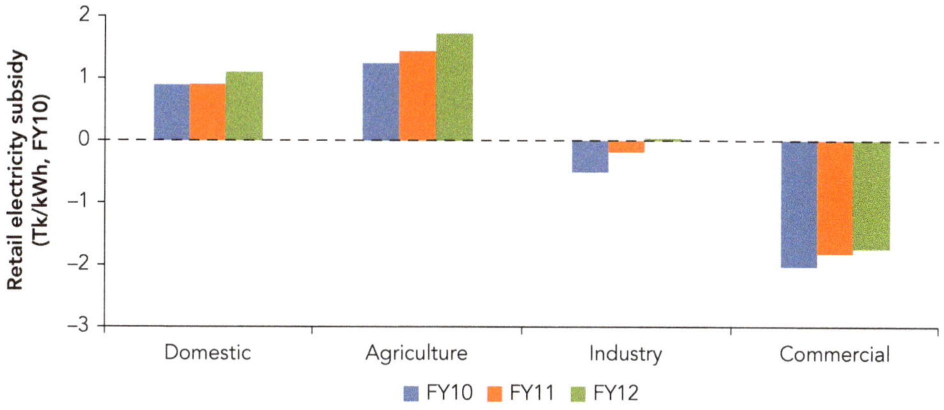

Source: Mujeri, Chowdhury, and Shahana (2013).
Note: FY = fiscal year; kWh = kilowatt-hour; Tk = Bangladesh taka.

The subsidies inflate demand for electricity, lead to greater dependence on imported fuel, and increase the fiscal burden on the government. They also limit the ability of BPDB to invest in expanding the electricity grid and improving the quality of electricity supply. Moreover, the subsidies are regressive. A World Bank study finds that households in the poorest quintile received only 9 percent of residential electricity subsidies in 2012, whereas those in the richest received 24 percent (Ahmed, Trimble, and Yoshida 2013).

Another distorting effect of electricity underpricing is that it encourages inefficient captive power generation. According to estimates by the Asian Development Bank, the cost of self-generation is about Tk 2–3 per kWh in 2014. By contrast, tariffs for grid electricity range from Tk 7 to Tk 11 per kWh for businesses (ADB 2014). The gas tariff for captive power generation has since been increased twice and by June 2017 was more than double its 2014 level. Self-generation is still likely to be less costly than grid electricity. For businesses, self-generation therefore makes economic sense. Indeed, captive power generation is the second-largest source of gas demand, accounting for 16 percent of gas consumption in fiscal 2016 (Petrobangla 2017). But providing gas to captive power plants could exacerbate the gas crisis because these plants are typically less efficient than utility-scale power stations in turning gas into electricity.

To quantify the welfare loss from price distortion, the analysis focuses on estimating the static deadweight loss from pricing below cost recovery levels—that is, removing subsidies would cause movement along rather than a shift of the supply and demand curves. At a weighted-average price of Tk 5.38 per kWh, the electricity supply falls short of demand by approximately 2.32 terawatt-hours (TWh) in fiscal 2016. The average wholesale purchase cost is Tk 5.90 per kWh, and the average transmission and distribution cost is estimated at Tk 0.82 per kWh (BPDB 2015). These differences imply subsidies to consumers of roughly 25 percent of the supply cost in fiscal 2016.

Deadweight loss arises when electricity is sold to consumers whose willingness to pay for electricity is lower than the level of the marginal cost of supply, resulting in a cost to taxpayers. On the basis of the estimated long-run supply and demand elasticities for electricity (see Box 3.1), the analysis projects the equilibrium price and quantity when subsidies are removed and the market clears. The predicted price at which there is no unmet demand is Tk 6.56 per kWh. At this price, demand declines by 4.8 percent. Both consumer and producer surpluses fall because of the lower transaction volume, but the savings in government subsidies more than offsets the loss. On the basis of this analysis, the total welfare cost of electricity subsidies is estimated at $867 million in fiscal 2016.

This analysis does not consider how removing subsidies could enhance BPDB's ability to maintain and improve the electricity supply, an effect that could increase consumers' willingness to pay for electricity (so that the supply and demand curves both shift downward). The results should therefore be considered an extremely conservative estimate of the regulatory cost of underpriced electricity.

SOCIAL: GAS WASTE LEADING TO POLLUTION FROM OIL USE

Inefficient electricity generation has resulted in waste in the use of gas and thus greater reliance on furnace oil and high-speed diesel for power generation, as discussed earlier about institutional distortions in the core segment. Besides being more expensive, liquid fuel pollutes more.

To estimate the associated social cost, the analysis uses a counterfactual scenario in which BPDB and rental power plants, in response to better incentive mechanisms, improve their fuel efficiency. The simulation focuses on the potential emissions savings from higher efficiency. The simulated improvement in efficiency is applied to each half-hour of the day. As in the previous exercise, plants use the gas saved each half-hour to produce more electricity during that period (while not exceeding their installed capacity). The additional production is applied first to reducing the unserved energy demand and then to replacing oil- and diesel-based generation. The avoided generation and emissions from oil- and diesel-based power plants are then computed.

The simulation results show that, after the efficiency improvement, generation based on liquid fuel would fall by 15 percent, or 1.36 TWh a year (Figure 3.22). The decline would lead to a corresponding reduction in carbon dioxide emissions of 340,000 tons a year. Using a shadow price of carbon dioxide emissions of $40 per ton, the avoided external costs would amount to $13.6 million a year. This estimate is a lower bound because it does not reflect the health benefits from reducing oil consumption.

FIGURE 3.22 Improving the efficiency of power generation would reduce the need for liquid fuel in Bangladesh

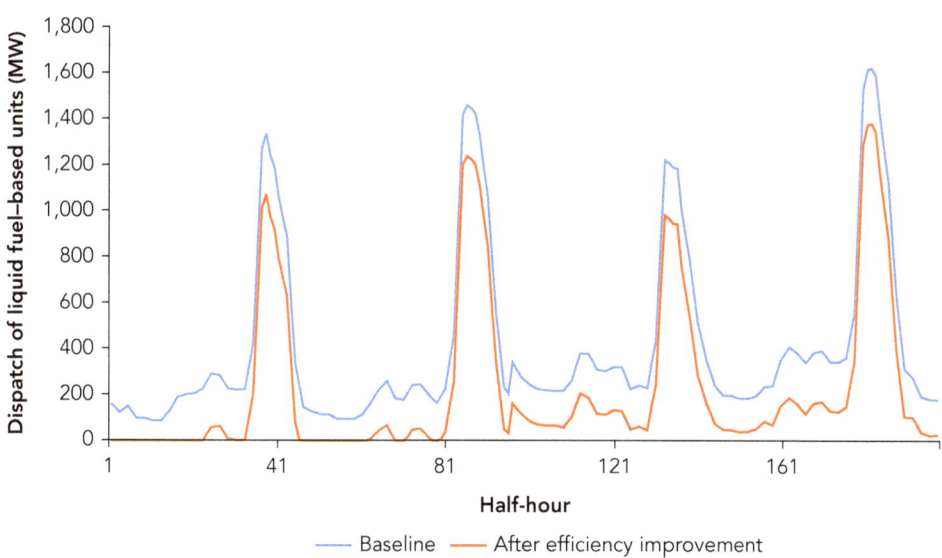

Source: Simulation based on daily reports, Bangladesh Power Development Board, January 1–December 31, 2014.
Note: For illustration, this figure shows data from January 1–4, 2014. The pattern remains the same for the year as a whole. MW = megawatt.

Downstream

Access to electricity, long recognized as a key driver of social and economic development, is lower in Bangladesh than in any other South Asian country. In the country's rural areas, less than 70 percent of households are connected to grid electricity. But, even for households and firms that are connected to the grid, the electricity supply could be intermittent because of technical failures and load shedding to deal with power shortages. Lack of access to a reliable supply limits economic opportunities, stifling growth and prosperity. According to a 2013 World Bank Enterprise Survey, more than half of business managers identified electricity as a major constraint on their operations.

Using detailed data from household and firm surveys in Bangladesh, this section quantifies the effects of electricity shortages on welfare loss for households, productivity loss for firms, and environmental costs.

INSTITUTIONAL: WELFARE LOSS FOR HOUSEHOLDS

The household-level analysis relies on data from two rounds of a World Bank–sponsored household survey conducted under the auspices of the Bangladesh Rural Electrification Board (REB). The first round, carried out in 2005, covered a nationally representative sample of 20,900 households from about 1,300 villages. A follow-up round in 2010 reinterviewed a randomly selected sample of 7,352 households from the original population. Both rounds of the survey include detailed questions on households' social and economic characteristics, income, expenditures, and energy use patterns. Although newer household survey data are available from other sources, the REB data set used in this analysis has two advantages. First, the panel nature of the household survey makes it possible to control for time-invariant household characteristics that may affect both the decision to adopt electricity and household economic and social outcomes. Second, the 2005 survey provides unique information on the quality of electricity services. Households were asked to report the daily average duration of electricity outages in hours. The analysis based on the REB data is used to identify the causal relationship between access to grid electricity and household welfare outcomes. The results are then used to estimate the cost of the lack of reliable access to electricity on households on the basis of the electrification rate in fiscal 2016.

Using the survey data, a simple average comparison between households with electricity and households without it provides suggestive evidence of the positive effects of electrification (Figure 3.23). On average, households that are connected to the grid consume less kerosene and have higher incomes and expenditures. Girls in households with electricity spend more time studying, and women in these households are more likely to own income-generating activities and have more decision-making power.

FIGURE 3.23 Households in Bangladesh with access to electricity had better welfare outcomes than households without access

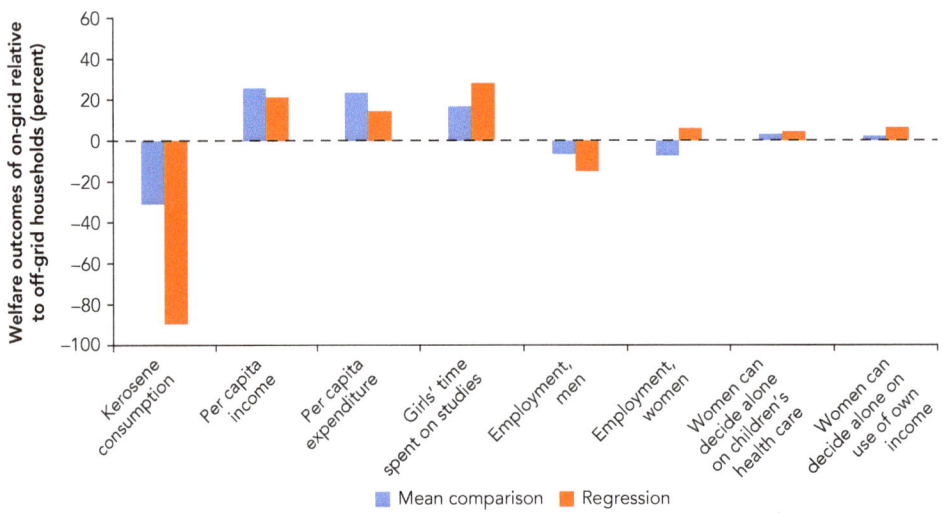

Source: Estimation based on Bangladesh Rural Electrification Board household surveys (2005 and 2010).
Note: Mean comparison refers to a simple average comparison without controlling for confounding factors. Regression refers to the estimated difference based on econometric analysis.

Observational evidence also indicates that households with more reliable access to electricity enjoy better economic outcomes (Samad and Zhang 2017). Five groups of households were sorted by the duration of the daily outages they face: up to 5 hours, 6–10 hours, 11–15 hours, 16–20 hours, and 21 hours or more. More hours of power outages are generally correlated with more kerosene consumption, and fewer hours are correlated with higher income and expenditure, more hours of study for children, greater labor force participation for both men and women, and greater empowerment of women. But, where electrified households face outages of 21 hours or more a day, there are almost no distinguishable differences between these households and those without electricity.

Correlation does not, however, necessarily imply causation. A government may target for electrification projects regions that are both poorer and growing more slowly. And, once electricity becomes available in a village, households that have better knowledge of its benefits or that can afford electrical appliances are more likely to adopt it. These preexisting differences may create differential trends in income growth and other outcomes even in the absence of electricity.

To address this potential selection bias, the analysis takes advantage of the fact that, even for households in the same village, the connection cost varies with proximity to the nearest electric pole: the closer the pole, the lower the cost. For a household located within 100 feet of the nearest pole, connection is free of charge. The typical connection cost is not low. According to a 2006 study, the average connection cost per household was Tk 2,800 ($40), or about 5 percent of the annual household expenditure (Mainuddin 2006). Free or low-cost connection therefore is a strong incentive to adopt grid electricity, but plausibly it does not otherwise affect economic or social outcomes for households in the same village. The analysis uses a binary variable measuring whether a household is located within 100 feet of an electric pole as a surrogate variable for access to electricity. There appears to be no selection bias for whose power is cut, because the data reveal that rich and poor households are equally affected by power outages, probably because load shedding is typically imposed on a wide geographic area. The analysis also controls for a range of observable household and village characteristics such as whether a household owns a solar home system. (For details on the methodology, see Samad and Zhang 2017.)

Results of the analysis show that electrification brings multiple benefits, but unreliability of supply can significantly reduce the size of those benefits (Figure 3.24). The first immediate benefit of gaining access to electricity is a reduction in kerosene consumption. Kerosene lamps are the primary source of lighting in households without electricity. Once households are connected to the grid, their kerosene consumption falls on average by 73 percent. But kerosene consumption rises with the duration of outages: on average, each one-hour increase in daily blackouts is associated with a 6 percent increase in kerosene use.

After gaining access to grid electricity, households also experience a significant increase in per capita income, 17 percent on average during 2005–10 (the sample period).

FIGURE 3.24 Households exposed to longer power outages had worse welfare outcomes in Bangladesh

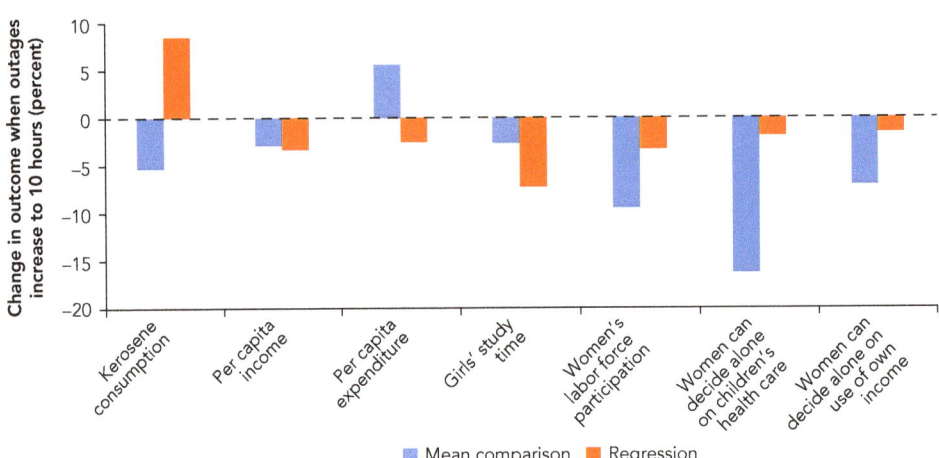

Source: Estimation based on Bangladesh Rural Electrification Board household survey (2005).
Note: Mean comparison refers to a simple average comparison without controlling for confounding factors. Regression refers to the estimated difference based on econometric analysis.

The growth in income seems to stem mostly from an increase in nonfarm income, suggesting that electricity may provide opportunities for more diversified economic activity in rural areas. Without electricity outages, nonfarm income would go up by 34 percent on average over the sample period (farm income is also positively correlated with electrification, but the relationship is not statistically significant). In other words, power outages dampen the positive effects of electrification on household income, especially for nonfarm income. Every one-hour increase in daily power outages is correlated with a reduction of 0.3 percent in total per capita income and 0.6 percent in per capita nonfarm income.

Corresponding with an increase in income, households that connect to the grid also increase their nonfood spending, by 21 percent on average over the sample period. As a result, total per capita spending increases by 12 percent. But every one-hour increase in daily power outages is associated with a 0.8 percent reduction in per capita nonfood spending.

Electrification can also help free up time for women and girls from domestic chores, allowing them to engage more in productive activities such as education, employment, and social participation. Indeed, access to grid electricity increases girls' study time and women's labor force participation, but it has no such effect on boys' study time or men's labor force participation. Power cuts again lessen the positive effects of electrification, with every one-hour increase in daily power outages reducing girls' daily study time by 0.11 hour.

Access to electricity also boosts women's empowerment. Women with access to electricity are more likely to decide alone on children's health care and on the use of their own income, probably because access to grid electricity helps increase women's economic empowerment through increased labor force participation. It also increases access to information: women in grid-connected households spend three times as much time watching television or listening to the radio as women in off-grid households.

All these benefits from expanding and improving the electricity supply are important. Not all can be quantified in monetary terms, but the potential gains in income growth alone are substantial. Assume that most unelectrified households are in rural areas and that their average annual household income was $2,002 in 2014 (Bangladesh Bureau of Statistics 2015). With estimated average income gains of 17 percent a year, the per household income gain from electrification is about $340 a year. Connecting the remaining unelectrified households to the grid would therefore increase income by $2.8 billion a year.

That said, the costs also have to be taken into account. Khandker, Barnes, and Samad (2012) estimate that the average cost of grid connection—which covers poles, lines, transformers, and related costs—is about $36 a year per household in a high-cost scenario. With the marginal cost of generation and distribution of electricity assumed to be about $0.09 per kWh in 2014 and average annual electricity consumption estimated at 680 kWh for rural households, the benefits outweigh the costs by a factor of 3.5. The net income gains from achieving universal access to electricity are estimated at $2.0 billion (1.2 percent of GDP) a year in Bangladesh.

Eliminating power outages would generate additional benefits. On the basis of an estimated income loss of 0.3 percent for each hour of daily power outage and a conservative estimation that households faced an average power outage of one hour a day in fiscal 2016, the additional benefits from improving the reliability of electricity supply is estimated at $260 million a year.

INSTITUTIONAL: PRODUCTIVITY LOSS FOR FIRMS

Because electricity is an essential input for most business processes, an unreliable supply can hurt firms' productivity. The analysis quantifies this effect by combining data on electricity shortages from BPDB with firm-level input-output information from industry surveys. Shortages are measured by the amount of load shedding (in megawatts), using data aggregated from BPDB daily generation reports for 2009–15. The firm-level data come from two sources. The first is the Survey of Manufacturing Industries conducted by the Bangladesh Bureau of Statistics. It covers total production, value of fixed assets, number of employees, consumption of raw materials, and energy expenditures. Although there have been six waves of this survey (fiscal 1995, 1997, 1999, 2001, 2005, and 2012), the analysis uses only the fiscal 2012 data because they can be matched with the shortage data. The 2012 survey covers 8,429 firms in nine manufacturing industries at the two-digit level of the International Standard Industrial Classification, including all large firms and a sample of micro-, small, and medium-size firms.

The second source of firm-level data is the World Bank Enterprise Survey, two rounds of which (2011 and 2013) overlap with the shortage data. This survey has a smaller sample size, but about 120 firms were surveyed in both years, allowing the analysis to explore within-firm variation to control for time-invariant characteristics unique to individual firms that may affect both a firm's demand for electricity and its performance.

One challenge in drawing causal inferences about the relationship between electricity shortages and firm outcomes is that unobservable factors are related to both shortages and productivity. For example, a region may implement policies that increase the productivity of firms, which in turn would increase the demand for electricity and worsen the shortages. A naive estimation of the effect that shortages have on firm outcomes could lead to an erroneous conclusion that shortages increase firm productivity (Allcott, Collard-Wexler, and O'Connell 2016; Grainger and Zhang 2017).

Variation in weather conditions provides an opportunity to avoid these biases. High temperatures increase firms' use of electricity, but plausibly such temperatures would not affect their productivity. (If high temperatures do affect productivity substantially through an effect on labor productivity, then the impact of power shortages would be overestimated. The analysis compares the estimation results with other estimates from the World Bank Enterprise Survey for a robustness check.) Following Fisher-Vanden, Mansur, and Wang (2015), the analysis computes the number of cooling degree days—the days on which the temperature exceeds 72°F—using temperature data provided by the National Climatic Data Center at the National Oceanic and Atmospheric Administration. The analysis calculates the daily average temperature for each region using hourly temperature data from 34 weather stations in Bangladesh, and then constructs the cooling degree days by year and region. The cooling degree days are used to predict the level of power shortages on the basis of the observed historical relationship between the temperature and actual shortage (used as a surrogate variable for both the cross-sectional data from the Survey of Manufacturing Industries and the panel data from the World Bank Enterprise Survey).

The analysis also controls for other relevant observable factors such as firms' sector classification and energy intensity. Firms' overall productivity (or total factor productivity) is estimated as the residual from a Cobb-Douglas production function in which input shares are allowed to vary from sector to sector. Because its measure of productivity is revenue-based (output is measured by sales value), the analysis captures the effects of shortages on both the quantity of production and the quality of output (through the impact on the price of products).

The analysis finds that a shortage of electricity leads to a significant reduction in firms' overall productivity. The effect is both statistically strong and economically nontrivial. Analysis based on cross-sectional data from the Survey of Manufacturing Industries suggests that a 10 percent shortage leads to a 3.1 percent reduction in a firm's total factor productivity; analysis based on panel data from the World Bank Enterprise Survey suggests that it causes a 4.1 percent reduction (Table 3.1). Notably, a simple regression analysis that does not address potential simultaneity bias suggests that electricity shortages are positively correlated with productivity. This finding shows the importance of using cooling degree days as an indirect measure of shortage.

The estimation results are lower than those self-reported by business managers. Responding to a question in the World Bank Enterprise Survey in 2013 when load shedding was about 11.3 percent of total demand, managers in manufacturing estimated that electricity shortages caused losses averaging 5.4 percent of annual sales; managers in the services sector reported losses averaging 6.0 percent of sales. Only limited data are available to gauge the effect of shortages on the services sector. But the results of the World Bank Enterprise Survey suggest that it is probably on the same order of magnitude (5.3 percent reduction in sales for a 10 percent power shortage) as in manufacturing (4.8 percent reduction in sales for a 10 percent power shortage)—see Table 3.1. Average load shedding in fiscal 2016 was about 2.2 percent of demand. Using the most conservative estimate of the impact of load shedding on firms' value added at 3.1 percent, along with 2016 data for Bangladesh on total value added in manufacturing ($37.7 billion) and services ($118.8 billion), the analysis estimates that for firms the total output loss associated with unreliable electricity is $1.1 billion (0.5 percent of GDP) a year.

This result is likely to underestimate the actual impact of power outages because the analysis uses changes in revenue as a proxy for changes in value added, ignoring the impact of outages on intermediate inputs. When electricity becomes scarce, firms may decide to outsource the production of energy-intensive intermediate goods rather than produce them in-house (Fisher-Vanden, Mansur, and Wang 2015). As a result, the cost of intermediate goods is likely to increase, further reducing the amount of value added.

TABLE 3.1 The estimated impact of electricity shortages on firms is similar across data sources and methodologies

Percent

	Data econometrically estimated		Data self-reported to World Bank Enterprise Survey	
	Survey of Manufacturing industry (2012, Bangladesh Bureau of Statistics)	Enterprise Survey, manufacturing (2013, World Bank)	Manufacturing	Services
Effect of 10 percent power shortage on productivity	−3.1	−4.1	−4.8	−5.3

Source: Estimation based on Survey of Manufacturing Industries in 2012; World Bank (2011, 2013).

SOCIAL: EMISSIONS FROM KEROSENE LIGHTING AND SELF-GENERATION

Without access to a reliable supply of electricity, households and businesses must rely on other sources of fuel such as kerosene to meet basic lighting needs. Tedsen (2013) estimates that almost 6.8 million kerosene lamps are being used by households and 1 million by businesses in Bangladesh. The use of kerosene as a lighting source has important health and environmental risks. Kerosene is a known cause of indoor air pollution, and many studies have reported a strong association between kerosene lighting and tuberculosis risk and respiratory infections. According to the World Health Organization, in 2012 about 85,000 deaths in Bangladesh were attributable to indoor air pollution (WHO 2012).

Precisely quantifying the health effects on households of lack of reliable electricity is difficult, however, because indoor air pollution in Bangladesh is also (and largely) attributable to the use of firewood and biomass for cooking. Electricity is not a preferred fuel for cooking even in urban areas, where electrification rates are estimated at more than 90 percent. Because of the difficulty in differentiating between emissions from lighting and emissions from cooking, the analysis does not estimate the health effects of kerosene lighting.

Kerosene lamps, as a source of black carbon emissions, also have substantial environmental costs. Black carbon is the second-largest source of climate warming after carbon dioxide. In 2005 residential lighting in Bangladesh produced an estimated 41.9 tons of black carbon, equivalent to 37,245 tons of carbon dioxide (Bangladesh Department of Environment 2014). On the basis of the difference in the access rate between 2005 and 2016, and assuming a shadow price for carbon dioxide emissions of $40 per ton, the environmental cost of black carbon emissions from kerosene-based residential lighting would be $0.6 million a year.

Unreliable access to grid electricity also has environmental consequences because it leads to greater use of fossil fuel for captive power generation. In Bangladesh, more than 60 percent of businesses own or share a generator. Because captive generators tend to be less efficient and closer to population centers than utility-scale power plants, they also leave a bigger environmental footprint. Almost all captive generators in Bangladesh use gas (IEA 2017). Their associated environmental effects are, however, not quantified because of lack of data on the efficiency of captive power generators.

Summarizing the Costs

Distortions in the power sector imposed a total economic cost of roughly $11.2 billion (5.0 percent of GDP) on Bangladesh's economy in fiscal 2016 (Table 3.2). The fiscal cost, which consists of electricity subsidies to consumers, was $0.33 billion (about 0.15 percent of GDP).

TABLE 3.2 Cost of power sector distortions in Bangladesh at a glance

Percent of GDP

| Type of cost | Upstream | Core | | | | Down-stream | Total |
		Generation	Dispatch	Transmission	Distribution		
Fiscal	0	0	0	0	0.15	0	0.15
Institutional	0.06	0.16	0.73	—	0.03	1.50	2.46
Regulatory	2.00	0	—	—	0.39	—	2.38
Social	0.16	0.01	—	—	—	0.003	0.16
Economic	2.21	0.16	0.73	—	0.41	1.48	5.01

Source: World Bank estimation.

Note: — = Not available. Estimation is for fiscal 2016.

The greatest source of waste occurs in the upstream gas sector. Selling gas at artificially low prices costs Bangladesh an estimated $4.5 billion (2.0 percent of GDP) a year, which could be spent promoting more sustainable long-term growth of the economy. The regulatory cost of gas underpricing could be even higher if its adverse effect on long-term growth were considered.

The second-largest source of distortion is households' and firms' lack of reliable access to electricity. It is estimated to cost the economy $3.3 billion (1.5 percent of GDP) a year. This cost consists of both the income forgone by the approximately 8.2 million households that still live without access to grid electricity and the revenue loss by firms that suffer from lower productivity and higher production costs as a result of electricity outages. Power shortages also negatively affect education, health, and women's empowerment. These effects are difficult to quantify and are not included in the estimation. The cost of lack of access to electricity could therefore be much higher than estimated here.

The third-largest source of distortion is out-of-merit dispatch of electricity. More expensive furnace oil–based power plants are often called on before gas plants are used. The cost of this inefficient dispatch is estimated at $1.65 billion (0.73 percent of GDP) a year (World Bank 2015b).

Other large costs stem from the social cost of excessive gas consumption, estimated at $355 million (0.16 percent of GDP) a year, and the inefficient allocation of gas, estimated at $130 million (0.06 percent of GDP) a year.

The analysis applies generally conservative assumptions throughout. Some of the analysis, including that on inefficient allocation of gas and inefficient dispatch of power plants, was based on data from 2014, the latest year for which data were available at the time. In addition, because of data limitations, it ignores some distortions, including transmission constraints, the social cost of electricity transmission and distribution, and the impact of electricity cross-subsidies on industry competitiveness. Overall, the estimate may represent a lower bound of the actual cost of power sector distortions.

References

ADB (Asian Development Bank). 2014. *Industrial Energy Efficiency Opportunities and Challenges in Bangladesh*. Manila, Philippines.

Ahmed, F., C. Trimble, and N. Yoshida. 2013. "The Transition from Underpricing Residential Electricity in Bangladesh: Fiscal and Distributional Impacts." World Bank Policy Note, Washington, DC.

Allcott, H., A. Collard-Wexler, and S. D. O'Connell. 2016. "How Do Electricity Shortages Affect Industry? Evidence from India." *American Economic Review* 106 (3): 587–624.

Bangladesh Bureau of Statistics. 2015. *Report on Education Household Survey 2014*. Ministry of Planning, Dhaka.

———. Various issues. *Statistical Year Book Bangladesh*. Ministry of Planning, Dhaka.

Bangladesh Department of Environment. 2014. *Bangladesh National Action Plan (NAP) for Reducing Short Lived Climate Pollutants (SLCPs)*. Dhaka.

BOGMC (Bangladesh Oil, Gas and Mineral Corporation). 2014. *Annual Report*. Dhaka.

———. 2015. *Annual Report*. Dhaka.

BPDB (Bangladesh Power Development Board). Various years. *Annual Report*. Dhaka.

Detsch, Jack. 2014. "Bangladesh: Asia's New Energy Power?" *The Diplomat*, November 14. https://thediplomat.com/2014/11/bangladesh-asias-new-energy-superpower/.

Fabrizio, Kira, Nancy L. Rose, and Catherine D. Wolfram. 2007. "Do Markets Reduce Costs? Assessing the Impact of Regulatory Restructuring on US Electric Generation Efficiency." *American Economic Review* 97 (4): 1250–77.

Fisher-Vanden, K., E. T. Mansur, and Q. J. Wang. 2015. "Electricity Shortages and Firm Productivity: Evidence from China's Industrial Firms." *Journal of Development Economics* 114: 172–88.

Grainger, C. A., and Fan Zhang. 2017. "The Impact of Electricity Shortages on Micro- and Small-Enterprises: Evidence from India." Background paper prepared for this report, World Bank, Washington, DC.

Hausman, Catherine, and Ryan Kellogg. 2015. "Welfare and Distributional Impact of Shale Gas." Brookings Papers on Economic Activity, Washington, DC, Spring.

Health Effects Institute. 2017. *State of Global Air: A Special Report on Global Exposure to Air Pollution and Its Disease Burden*. Boston.

IEA (International Energy Agency). 2017. World Energy Balance and Statistics Database. https://www.iea.org/statistics/relateddatabases/worldenergystatisticsandbalances/.

IRENA (International Renewable Energy Agency). 2017. *Renewable Energy and Jobs Annual Review*. Vienna.

Khan, Muhammad Arshad. 2015. "Modelling and Forecasting the Demand for Natural Gas in Pakistan." *Renewable and Sustainable Energy Reviews* 49: 1145–59.

Khandker, Shahidur, Douglas Barnes, and Hussain Samad. 2012. "The Welfare Impacts of Rural Electrification in Bangladesh." *Energy Journal* 33 (1): 199–218.

Mainuddin, K. 2006. "Access, Willingness to Pay and Affordability of Electricity." Bangladesh Centre for Advanced Studies (BCAS), Dhaka.

Ministry of Finance, Bangladesh. 2016. *Bangladesh Economic Survey 2015–2016*. Dhaka.

Mujeri, Mustafa, Tahreen Chowdhury, and Siban Shahana. 2013. *Energy Subsidies in Bangladesh: A Profile of Groups Vulnerable to Reform.* Winnipeg, Manitoba: International Institute for Sustainable Development.

Nikolakakis, Thomas, Deb Chattopadhyay, and Morgan Bazilian. 2017. "A Review of Renewable Investment and Power System Operational Issues in Bangladesh." *Renewable and Sustainable Energy Reviews* 68: 650–58.

Parry, Ian, Dirk Hein, Eliza Lis, and Shanjun Li. 2014. *Getting Energy Prices Right: From Principle to Practice.* Washington, DC: International Monetary Fund.

Petrobangla. Various years. *Annual Report.* Dhaka.

Rahman, Mohammad Azizur. 2016. "Bangladesh Launches Fresh Bidding Round for 3 Off-Shore Gas Blocks." S&P Global Platts, October 4. https://www.platts.com/latest-news/natural -gas/dhaka/bangladesh-launches-fresh-bidding-round-for-3-27681148.

Rasel, Aminur Rahman. 2014. "Conoco Phillips Pulls out of Deep-Sea Blocks 10, 11." *Dhaka Tribune*, October 26. http://www.dhakatribune.com/bangladesh/2014/10/26/conocophil lips-pulls-out-of-deep-sea-blocks-10-11/.

———. 2017. "Government Ponders New PCSs for Oil and Gas Exploration." *Dhaka Tribune*, August 27. http://www.dhakatribune.com/bangladesh/power-energy/2017/08/27/government -ponders-new-pscs-oil-gas-exploration/.

Reuters. 2013. "Santos Closes Down Bangladesh's Only Offshore Gas Field." October 1. https:// uk.reuters.com/article/bangladesh-offshore-energy/santos-closes-down-bangladeshs -only-offshore-gas-field-idUKL4N0HR2S420131001.

Samad, Hussain, and Fan Zhang. 2017. "Heterogeneous Effects of Rural Electrification: Evidence from Bangladesh." Policy Research Working Paper 8102, World Bank, Washington, DC.

Schwab, Klaus. 2018. *The Global Competitiveness Report 2017–2018.* Geneva: World Economic Forum.

Tedsen, E. 2013. *Black Carbon Emissions from Kerosene Lamps: Potential for a New CCAC Initiative.* Berlin: Ecologic Institute.

WHO (World Health Organization). 2012. Global Health Observatory Data Repository. http:// apps.who.int/gho/data/node.main.BODHOUSEHOLDAIRDTHS?lang=en.

World Bank. Various years. World Bank Global Economic Monitor Commodities (database). Washington DC. http://databank.worldbank.org/data/reports.aspx?source=global-economic -monitor-commodities.

———. 2013. Enterprise Survey. Washington, DC.

———. 2015a. *Bangladesh Development Update, April 2015.* Dhaka.

———. 2015b. "A Review of Renewable Investment and Power System Operational Issues in Bangladesh." Also published in 2017 in *Renewable and Sustainable Energy Reviews* 650–58.

———. 2016. *South Asia Economic Focus Spring 2016.* Washington DC.

Zhang, Fan. 2007. "Does Electricity Restructuring Work? Evidence from the U.S. Nuclear Energy Industry." *Journal of Industrial Economics* 55 (3): 397–418.

———. 2015. "Energy Price Reform and Household Welfare: The Case of Turkey." *Energy Journal* 36 (2): 71–95.

CHAPTER 4

India

India has made enormous progress in expanding household access to electricity and reducing power shortages over the last few years. In 2018 India achieved 100 percent village electrification. More than 115 million people have gained access to electricity since 2013, increasing the share of population with access to electricity from less than 80 percent in 2013 to 86 percent in 2017. With additional 40 million households targeted under the rural electrification scheme Saubhagya, the Government plans to provide universal access to electricity to all households by December 2018. Generation capacity has also increased substantially. The total installed generation capacity more than doubled over the past decade, from 154.7 gigawatts (GW) in fiscal 2007 to 345.5 GW in 2018. India is now the world's third-largest producer of electricity, after China and the United States. Thanks to the significant capacity added and the lower than expected demand growth, power shortages have been significantly reduced. The peak demand shortage declined from 11 percent in fiscal 2012 to 0.9 percent in fiscal 2018, while the average demand shortage fell from 8.5 percent to 0.4 percent during the same period (Figure 4.1). The Central Electricity Authority predicts that India is likely to become a power surplus country in fiscal 2019 (CEA 2018).

In recent years, India has also become one of the world's leading countries in renewable energy development. Its cumulative installed capacity of renewable energy reached 70 GW in 2018, about 50 percent of which had been installed since May 2014. The increase in solar and wind energy since 2014 has been particularly large. India now ranks fourth in the world in wind power–based capacity (after China, the United States, and Germany) and sixth in solar-based capacity (Press Information Bureau 2017). Through various initiatives and incentive programs, the government of India plans to add 227 GW of renewable energy capacity by the end of 2022: 114 GW from solar, 67 GW from wind, 31 GW from floating solar and offshore wind, 10 GW from biomass, and 5 GW from small hydro.

India has also taken important steps to improve energy efficiency. Under the Unnat Jyoti by Affordable LEDs for All (UJALA) scheme, launched in 2015, 240 million energy

FIGURE 4.1 Electricity shortages declined between 2012 and 2017 in India

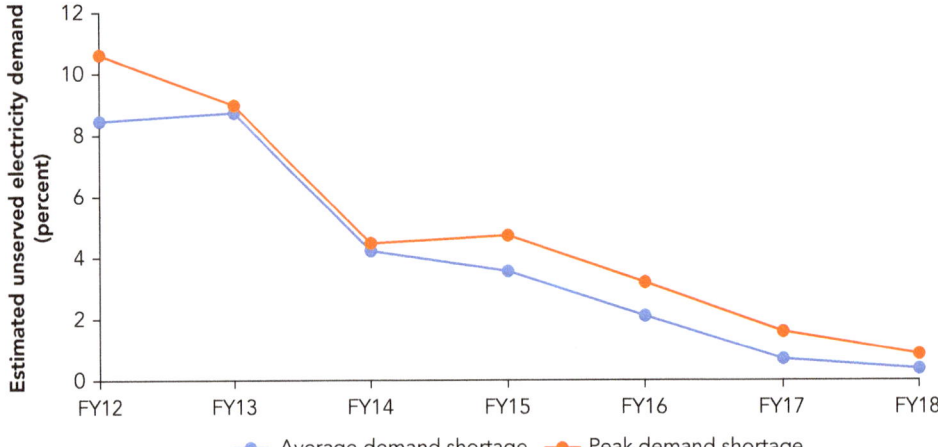

Source: Monthly reports, National Load Dispatch Center, accessed through Indiastat.
Note: FY = fiscal year.

efficient LED bulbs and 1 million energy efficient fans have been distributed to consumers as of July 2017. Another energy efficiency program, Perform Achieve and Trade (PAT), set targets for energy savings for energy-intensive sectors, including power generation. As of now, 208 thermal power plants have participated in the scheme and taken measures to achieve energy efficiency performance standards.

Notwithstanding this remarkable progress, India still faces an enormous need to meet the growing demand for electricity. Because of India's growing population, rapid urbanization, and economy that is expected to grow at an average rate of almost 7 percent per year, the International Energy Agency projects that electricity demand in India will almost triple between 2018 and 2040 (IEA 2017c). India also needs to further expand the access of households to electricity and improve the quality of electricity services. As of October 2017, about 178 million people were still living off-grid. Although the power deficit has been substantially reduced over the last few years, the reliability of electricity is still low compared with the international standard: the 2018 *Global Competitiveness Report* ranks India 80th among 137 economies in the reliability of its electricity supply (Schwab 2018).

Air pollution from fossil fuel–based power generation poses another daunting challenge. Despite ambitious programs to promote renewable energy development, India still uses coal to generate 75 percent of its electricity. Meanwhile, industries produce their own "captive" power, much of it from coal and small diesel generators.

Burning coal and diesel releases toxic pollutants. The most harmful is the fine particulate matter ($PM_{2.5}$) that can be inhaled deep into the lungs, causing illness and premature death. A recent global study finds that India's mortality rate associated with exposure to $PM_{2.5}$ is among the highest in the world—more than 1 million deaths and a loss of 29.6 million years of healthy life in 2015 (Health Effects Institute 2017).

About 7.6 percent of $PM_{2.5}$ emissions were attributable to power generation, 7.7 percent to industry coal combustion, and 2.0 percent to distributed diesel generators (Global Burden of Disease MAPs Working Group 2018).

Increasing access to reliable electricity is imperative for improving living standards in India and ensuring the success of initiatives such as Make in India, Skill India, and Digital India. This chapter identifies the main institutional, regulatory, and social distortions of power supply in India and quantifies their impacts. The results reveal that implementing comprehensive energy sector reform that targets inefficiencies at different stages of the power supply could boost supply while also limiting potentially harmful emissions.

Upstream

Coal is India's dominant source of energy. In 2015 it fueled 75 percent of the country's electricity production, the sixth-highest share globally (IEA 2017b). Since 1970, the amount of electricity produced from coal has grown markedly, whereas the amount generated from other sources has increased only marginally (Figure 4.2). With this growth has come a big increase in the share of the country's coal used for electricity generation—from 49 percent in 1971 to 75 percent in 2015.

India's coal reserves are estimated at 306 billion tons, the fourth-largest in the world. In 2016 India produced 692.4 million tons of coal, the second-largest amount in the world after China. Its coal production has gradually increased in recent decades, rising

FIGURE 4.2 Coal plays a dominant role in electricity generation in India

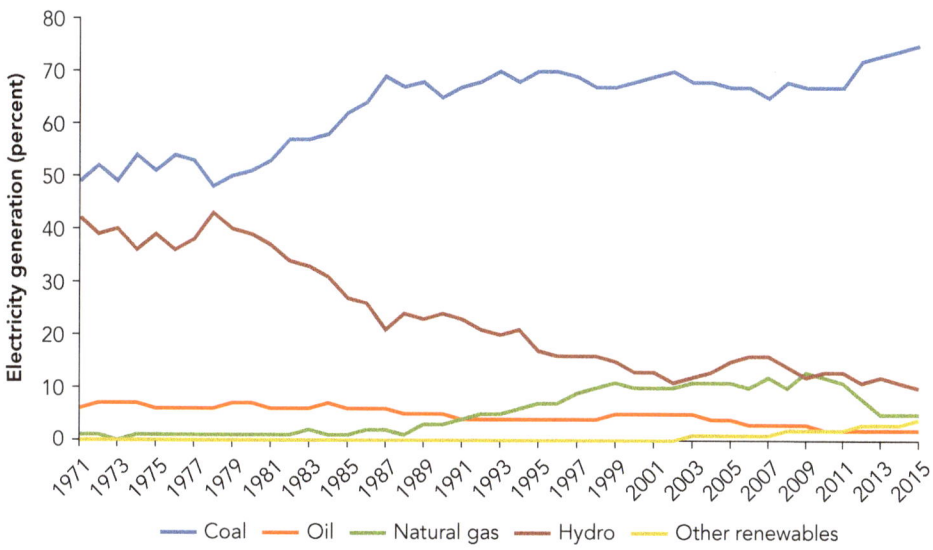

Source: IEA (2017c).

from 73 million tons in fiscal 1971 to 639 million in fiscal 2016. But growth has been sluggish compared with the increase in demand. The domestic coal supply fell 14 percent short of the requirements of power plants in fiscal 2016—the total gap between demand and indigenous supply in the power sector surged from 33 million tons in fiscal 2009 to 62 million tons in fiscal 2016 (Figure 4.3).

The shortage of domestic coal increased dependence on thermal coal imports, which rose from 9 million tons in fiscal 2000 to 156 million tons in fiscal 2016. Although imported coal could be less expensive than domestic coal in coastal areas after adjusting for transportation costs, in fiscal 2016 it was twice as expensive as domestic coal on average, even after taking into account differences in heat content. The greater dependence on imported coal has therefore led to higher generating costs. During this period, many new plants were designed and built to run only on imports. These plants were locked into foreign coal even during the recent surge in the domestic coal supply.

Coal shortages are exacerbated by rail transport constraints. India's coal reserves are located mostly in its eastern states, whereas coal-fired power plants are scattered across the country. Because of this geographic mismatch, coal must be hauled long distances, mostly by rail. Constraints in rail capacity have led to major bottlenecks, leaving millions of tons of coal stranded at mines (EIA 2015).

The coal shortfall has had a drastic effect on electricity generation. According to India's Central Electricity Regulatory Commission, lack of coal led to the stranding of about 48 GW of generation capacity in fiscal 2014, equivalent to 15 percent of the entire coal fleet (CERC 2015).

FIGURE 4.3 The shortage of coal for power generation increased significantly between fiscal 2009 and fiscal 2016 in India

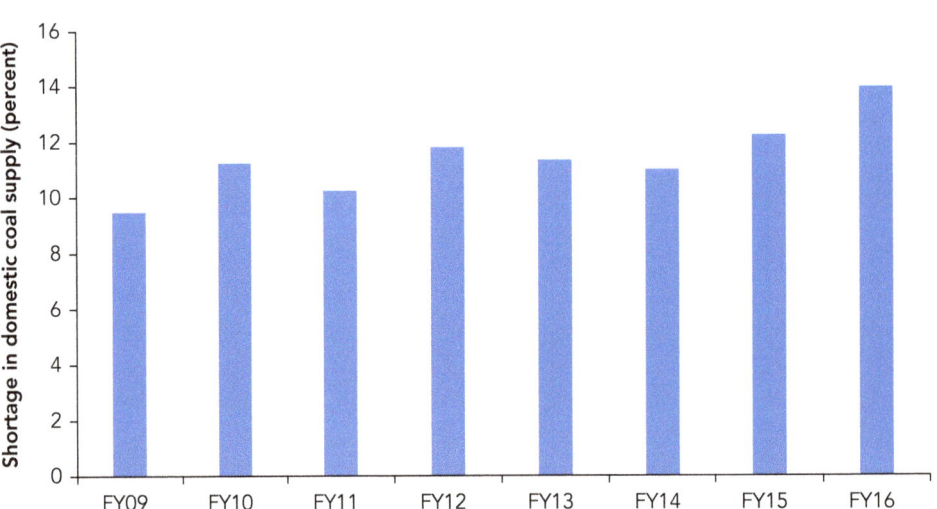

Source: Ministry of Coal, India, *Provisional Coal Statistics*, accessed through Indiastat.
Note: FY = fiscal year.

Insufficient investment, inadequate technology, and unproductive labor are among the factors underlying the coal supply shortfall. Meanwhile, the underpricing of coal for power generation artificially inflates demand, and the allocation of coal does not prioritize its efficient utilization. This section discusses these distortions.

INSTITUTIONAL: UNPRODUCTIVE MINING AND PRIVILEGED ACCESS

India nationalized coal mining in 1973. Coal India Limited (CIL), the largest publicly owned company in the country, tightly regulates the industry and sets the price for all coal producers. With its subsidiaries, CIL is also the world's largest coal company. During fiscal 2016, it produced 537 million tons of coal, accounting for more than 80 percent of domestic production.

The rest of India's domestic coal supply is provided by Singareni Collieries Company Limited (SCCL), another publicly owned mining company, and privately owned captive mines (Figure 4.4). To meet the growing demand for coal, the government allows private companies, such as power plants and iron and steel producers, to own coal mines for their own use. Coal produced from these captive mines cannot be sold on the open market. Only in 2018 did the government open the sector for private commercial mining.

Economists argue that lack of competition stifles incentives for innovation and productivity growth. This seems to be true for India's coal mining industry. Although advances in automation have made mineral extraction increasingly safe and efficient around the world, in India coal mining remains largely manual.

FIGURE 4.4 State-owned mining companies dominate the coal market in India

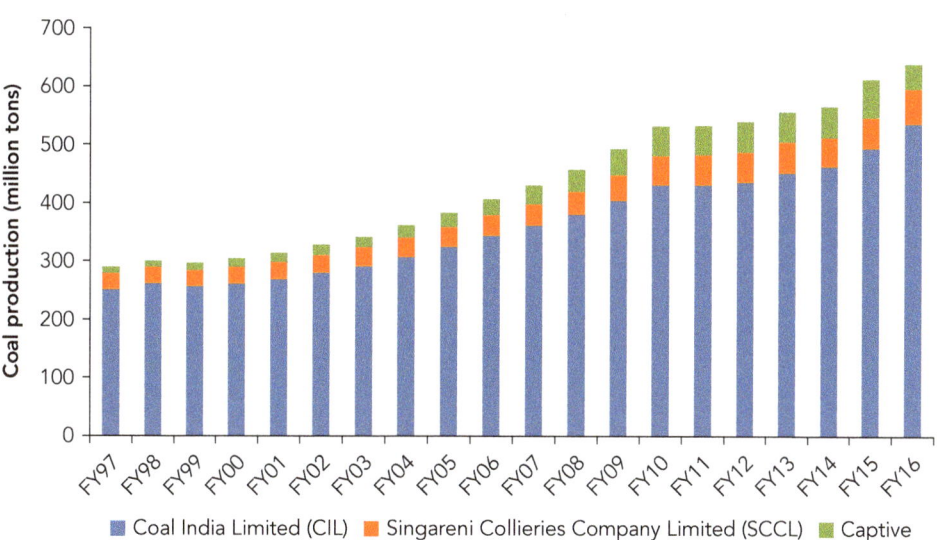

Source: Ministry of Coal, India, *Provisional Coal Statistics*, accessed through Indiastat.
Note: FY = fiscal year.

Low levels of mechanization particularly affect the production rates of underground (deep) mining because geological conditions worsen at deeper depths. In most of the world, the most common mass production technologies for underground mining are the longwall system and continuous mining. Extraction by these technologies is an almost continuous operation involving the use of self-advancing roof supports, a sophisticated coal-cutting machine, and a paralleling conveyor that automatically transports coal out of the mine. In the United States, the longwall system and continuous mining accounted for more than 99 percent of underground coal production in 2015 (EIA 2016). In India, by contrast, they accounted for only about 11.5 percent of underground output in fiscal 2016 (CIL 2016). Most underground mines still rely on traditional mining methods, often involving men digging, extracting, and hand loading coal on carts.

Government coal statistics reveal that capital investment in underground mines (as proxied by increases in machine horsepower) was almost nonexistent between 1993 and 2009 (a period for which data exist) and that labor productivity (as measured by output per labor year) remained stagnant (Ministry of Coal, India 2010). In fiscal 2016, average output per labor shift was 0.79 tons at CIL underground mines and 1.25 tons at SCCL underground mines. By contrast, in the United States the productivity of underground coal mining is equivalent to about 25 tons per labor shift, or more than 20 times greater (EIA 2016).

In other large coal-producing countries, underground coal accounts for a production share that is largely commensurate with the level of underground recoverable reserves. In China, for example, underground coal represents 92.5 percent of recoverable reserves and about 95.0 percent of production; in the United States, the corresponding shares are 38.9 percent and 34.0 percent (EIA 2016). In India, however, with at least 41.5 percent of recoverable coal reserves located at depths greater than 300 meters, the production share of underground coal dropped to only 7.3 percent in fiscal 2016 (down from 24.5 percent in fiscal 1997). With the failure to extract coal from deeper seams, opencast (surface) mining became the main technology for the industry. Its share of output rose from almost zero in 1951 to more than 90 percent in fiscal 2016.

Labor productivity in opencast mining increased substantially in India over the last few decades, but it remains much lower than in many other coal-producing countries. In fiscal 2015, the average output per labor shift in opencast mines was 15.3 tons for CIL and 13.8 tons for SCCL, or roughly one-fifth the level in Australia or the United States (EIA 2016; TERI 2013).

Coal deposits in India are not continuous, which has partially contributed to poor labor productivity. One could also argue that in India, because labor is relatively inexpensive compared with other inputs, mining is naturally more labor-intensive, and labor productivity therefore is lower than in other countries. But the productivity trends for capital and material are also discouraging. In underground mines, output per horsepower of machinery fell by 53.7 percent between 1996 and 2014 (Figure 4.5). All this points again to the way in which inadequate technology can lead to loss of productivity.

FIGURE 4.5 The productivity of machinery in underground coal mines declined
between 1996 and 2014 in India

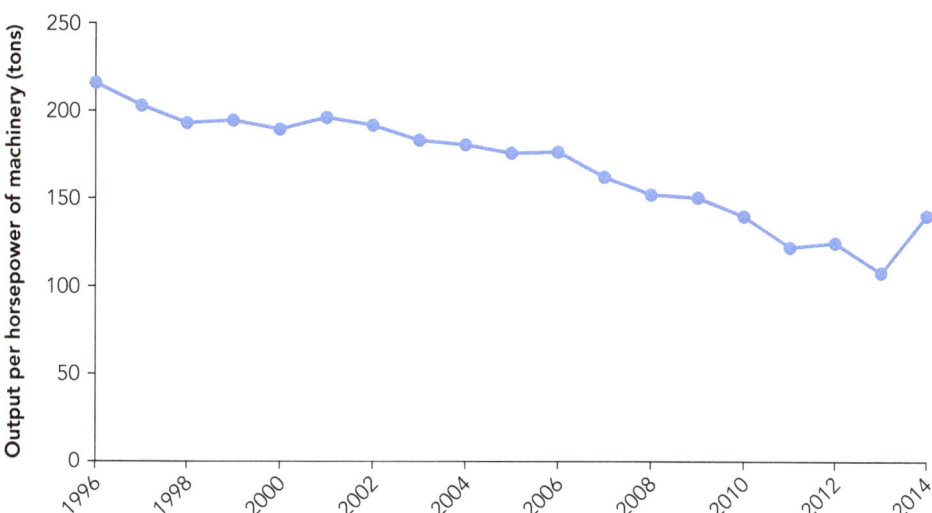

Source: Directorate General of Mines Safety (2014).

Misallocation of various types of resources compounds the low productivity in coal mining. Opencast mines have much higher labor productivity than underground mines, but underground mines, which contribute less than 7 percent of coal output, nevertheless commanded half the workforce in coal mining in 2014 (Figure 4.6). Labor productivity also varies greatly across mines of different types, suggesting the potential for increasing total output by reallocating labor from lower- to higher-productivity mines. Wages should reflect labor productivity, but their distribution across states is much narrower than the distribution of labor productivity, according to fiscal 2009 data (Figure 4.7), suggesting that wages have not played a big role in the allocation of labor across mines.

Scarce coal resources are also misallocated. Coalfields for captive mining used to be allocated through administrative orders. In 2014, however, the Supreme Court of India decided that the process was arbitrary and lacked transparency. Some 214 coal block licenses were revoked and set to be redistributed by competitive auction. Government-owned companies were exempt from competitive bidding, however; they were eligible instead for guaranteed allotment under a separate window. Thus among 89 recently reallocated coalfields, 41 were awarded to public power plants with no bidding process. Public power plants also receive priority in receiving CIL-produced coal. And, although all public power plants are linked to designated mines under long-term fuel supply agreements, private power plants owned by independent power producers (IPPs) often need to purchase coal on the spot market. When there are coal shortages, favoring

FIGURE 4.6 The labor productivity of opencast coal mines is much higher than that of underground coal mines in India

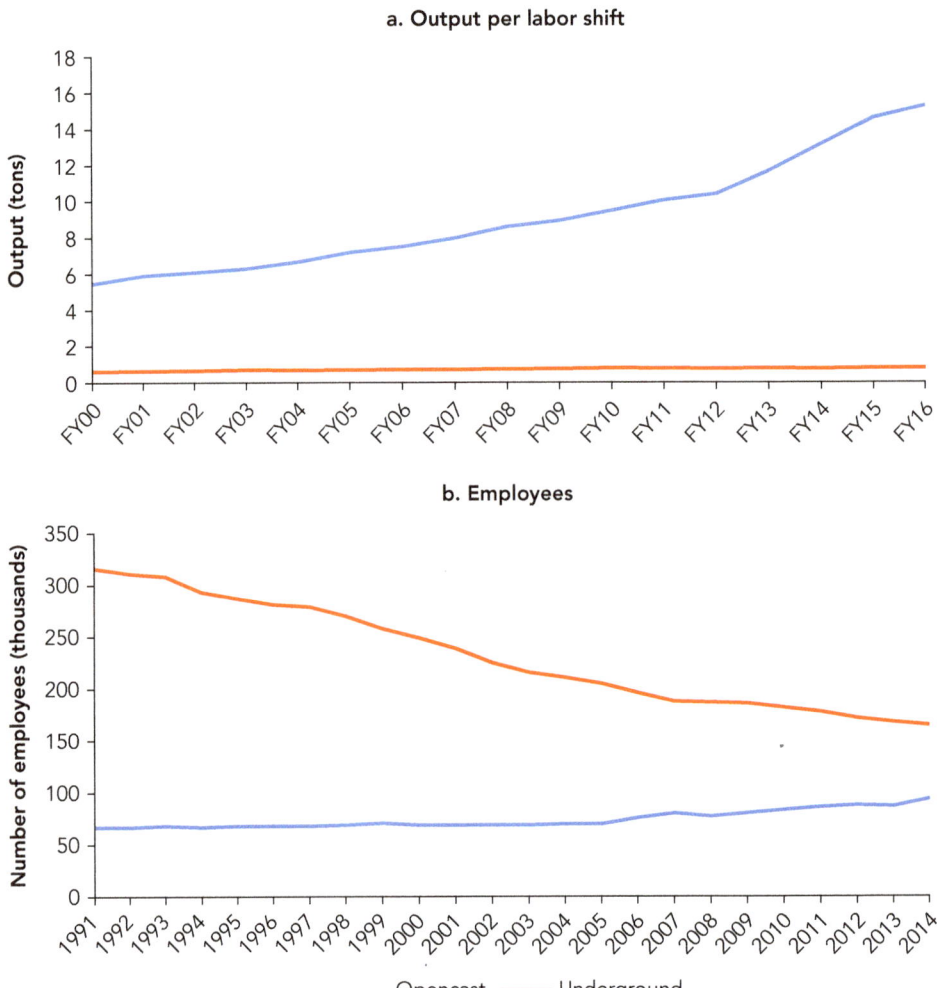

a. Output per labor shift

b. Employees

Opencast ——— Underground

Source: Ministry of Coal, India, *Provisional Coal Statistics*, accessed through Indiastat; Ministry of Labor and Employment, India (2016).
Note: FY = fiscal year.

public power plants at the expense of private ones will result in less electricity produced per unit of coal because public power plants are generally less efficient (see the section of this chapter on inefficient government-owned power plants).

Lack of data on labor allocation across mines and on coal shortages at different plants rules out an exercise quantifying the cost of resource misallocation. Analysis of the cost of upstream institutional distortions therefore focuses only on unproductive underground mining, and so the results are likely to be a lower-bound estimate.

FIGURE 4.7 **The dispersion of wages is much narrower than the dispersion of labor productivity in India**

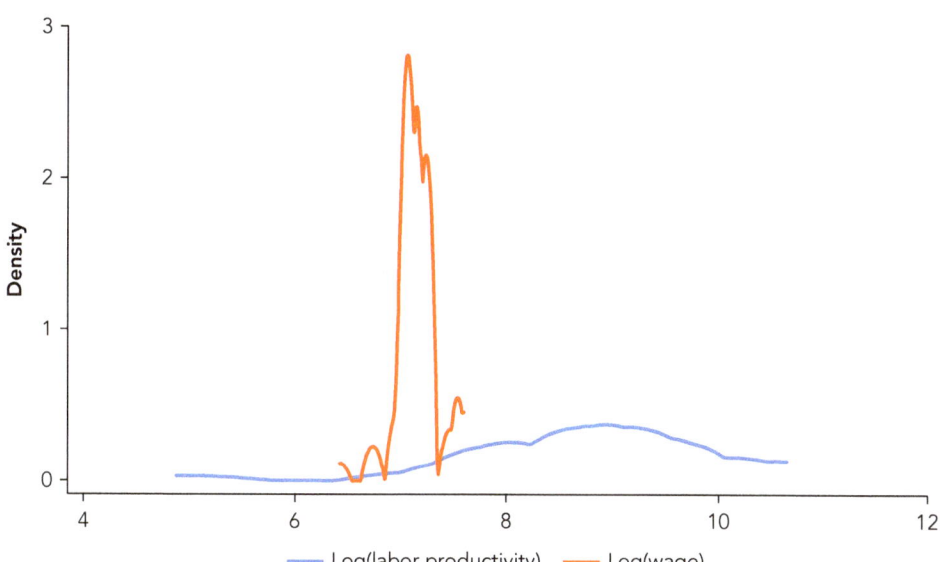

Source: Based on Ministry of Coal, India (2010).
Note: Labor productivity is measured by output (tons) per labor shift. Wage is wage index. Labor productivity and wage are state mean values for fiscal 2009.

To estimate the cost of unproductive underground mining, the analysis considers a counterfactual in which all unmechanized underground mines adopt automation in the form of longwall systems. After SCCL introduced a longwall set at Adriyala in 2011, the mine's average output per labor year increased to 550 tons in fiscal 2014. By comparison, the average output per labor year for unmechanized underground mines is estimated at 361 tons. Because in India 11.5 percent of output is from mines equipped with longwall systems and continuous mining, the analysis makes a conservative assumption that the same share (11.5 percent) of underground miners work in mechanized mines. It is therefore estimated that the productivity boost resulting from adoption of longwall systems across India would increase output by 28 million tons a year. This estimate is conservative compared with the conclusion of a study by the Australian Department of Industry and Science (2015), which estimates that the conventional mining method used by most CIL underground mines leaves behind about 40 million tons of coal a year in these mines. The study, citing other literature, estimates that CIL could extract 70 percent more coal from existing and future operations by introducing the longwall system.

Introducing more advanced mining technology would shift the supply curve of coal to the right. The new supply curve can be estimated on the basis of the scale of the potential output increase just described. Using the estimated price elasticities of supply and demand for coal in India (Box 4.1), the analysis then projects the new equilibrium quantity at the current notified coal price (after tax) for power producers. In fiscal 2016, imported thermal coal amounted to 156 million tons. More efficient production would avoid to some extent the higher costs of imported coal. The consumer surplus would increase by $742 million a year, and because of greater outputs the producer surplus would increase by an estimated $436 million a year. The benefit from upgrading the mining technology in underground mines is therefore estimated to be at least $1.18 billion (0.06 percent of the gross domestic product, GDP) a year.

Looking forward, efforts toward commercial mining can be expected to improve competition and increase efficiency, which would allow the infusion of advanced coal mining technologies in the sector. This will create more direct and indirect employment in coal bearing areas and will have an impact on economic development of these regions. Coal-bearing states would also directly benefit from revenue from the auction of coal mines. These benefits are potentially huge but are not quantified in the analysis.

BOX 4.1 Estimating the price elasticity of supply and demand for coal in India

This analysis estimates the price elasticity of coal supply and demand for the power sector in India, using income and price as the main determinants of supply and demand. The estimation of supply elasticity is based on annual data for domestic thermal coal production, consumption of coal for electricity generation, the estimated coal shortage in the power sector, the real average thermal coal price index, and the real per capita GDP over the period 1990–2014. Coal production, consumption, and shortage data are from various editions of *Provisional Coal Statistics*, published by India's Ministry of Coal, and the price index data are from the website of India's Ministry of Commerce and Industry. Real per capita GDP data are from the World Bank's World Development Indicators database.

To address the potential endogeneity of price, the contemporaneous coal price is instrumented by the first lag of the average mining wage and the oil price in estimating the supply curve. Wages and oil prices are two main drivers of the cost of coal mining in India because mining is extremely labor-intensive and oil is used as a fuel for earth-moving equipment and as an input for certain explosives. Wage data are obtained from various editions of the annual report of the Directorate General of Mines Safety, Ministry of Labor and Employment, accessed through Indiastat. The contemporaneous coal price is instrumented by its first two lags and the first lag of the price of substituting fuel (gas) in estimating the demand curve. A test is performed for the existence of long-run cointegration, and an error correction model is estimated to obtain both short- and long-run elasticities. (For details on the methodology, see Appendix A.)

The estimated long-run supply elasticity is 0.45, and the estimated long-run demand elasticity is −0.20, meaning that in the long run a 1 percent increase in the domestic price for coal would increase supply by 0.45 percent and reduce the power sector's demand for coal by 0.20 percent. Short-term coefficients on prices are 0.16 for the supply curve

box continues next page

BOX 4.1 **Estimating the price elasticity of supply and demand for coal in India** *(continued)*

and −0.16 for the demand curve. As expected, the short-term elasticities are smaller than the long-run estimates (in absolute terms). By contrast, analysis ignoring potential price endogeneity produces counterintuitive signs of price elasticities.

A few other studies estimate supply and demand elasticities for coal using econometric techniques. Kulshreshtha and Parikh (2000) find that the price elasticity of demand for coal for power generation in India is −0.125. According to Dahl (1993), the price elasticity of demand for coal in the United States is −0.4 in the short term and between −0.70 and −0.90 in the long run. Burke and Liao (2015) reveal that the demand elasticity in China is −0.20 in the short term and −0.40 in the long run. And Lawrence and Nehring (2015) discover that the elasticity of coal supply ranges from 0.12 to 0.30 in Australia and from 0.62 to 0.73 in the United States.

The estimated price elasticities for India are smaller than those found for Australia, China, and the United States. With larger elasticities, the economic cost of distortions would be even higher.

REGULATORY: UNDERPRICED COAL FOR POWER GENERATION

The pricing mechanism for coal in India has changed substantially since 2000. The government has deregulated the price of coal, switched from a 7-grade to a 17-grade pricing system, and introduced an environmental tax on coal production and imports. However, taking the productivity of CIL and SCCL as given, the price of coal, especially for power generation, remains heavily subsidized. On the basis of different benchmark prices, underpricing of coal is estimated to range from 17 percent to 100 percent of the existing tariff.

Historically, the central government fully controlled the prices of different types and grades of coal. In 1996, however, it began to gradually deregulate the price of coal, fully completing the process in 2000. Coal pricing now depends entirely on the price notified by CIL. In practice, however, CIL sets the price in consultation with the government. The policy influence is most evident in the differential pricing for the so-called regulated sectors: power, fertilizer, and defense. These sectors receive a discounted price that is 17 percent lower on average than the price charged to unregulated sectors (CIL 2018).

The shift from an administered to a deregulated price regime has not resulted in significant changes in overall price levels. The real wholesale price index for thermal coal remains flat and has even declined slightly since price deregulation in 2000 (Figure 4.8).

Reform of the grading system for coal has a greater impact on bringing the domestic coal price in line with international prices. Compared with imported coal, Indian coal is poor in heat content and rich in ash content. Its quality also varies widely. Until 2012, coal was classified into seven grades (A–G) on the basis of useful heat value. The band for each grade ranges in width from 600 kilocalories (kcal) per kilogram for the highest grade

FIGURE 4.8 Deregulation did not increase the wholesale price for thermal coal in India

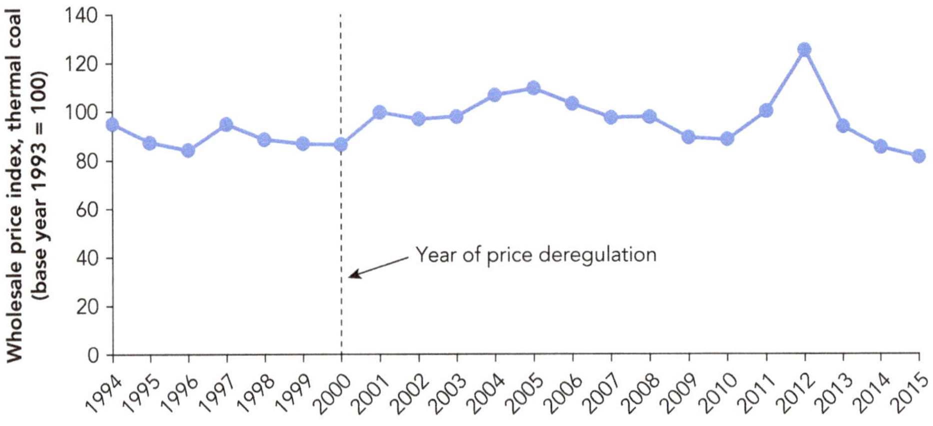

Source: Ministry of Commerce and Industry, India, Wholesale Price Index.

to 1,100 for the lowest. The wide-band pricing provides little incentive for coal producers to improve the quality of coal through measures such as coal washing (beneficiation)
to reduce the ash content. By some estimates, it has also contributed to the practice of
pricing domestic coal 50–65 percent lower than the international price on an energy-
equivalent basis (Tiwari, Bhattacharya, and Raghav 2015).

At the beginning of 2012, CIL announced a switch to a grading system based on
gross calorific value (GCV). It classifies coal into 17 grades (G1–G17) at a narrower
uniform interval of 300 kcal per kilogram. The new grading system is more consistent
with the standard followed internationally and would help bring domestic coal prices
closer to international rates. The new pricing policy led to a 5–12 percent increase in
the prices of different grades of coal.

India is among the few countries that have imposed an environmental tax on coal consumption. The Clean Environment Cess was introduced in fiscal 2010 at 50 rupees (Rs) per
ton of coal. It was increased to Rs 100 per ton in fiscal 2014, Rs 200 per ton in fiscal 2015, and
Rs 400 ($6) per ton in fiscal 2016. The tax raised the price of coal by 21 percent for the high-
est grade and 85 percent for the lowest one in 2016. It has helped internalize the social costs
of coal consumption, although at its current rate it offsets less than 3 percent of the health
and environmental damages caused by coal-based power generation (for more details, see
the section in this chapter on emissions, disease, and accidents from coal).

Despite the various pricing reforms, the price of coal for power generation has been
kept low as a way to subsidize electricity in India. Although overall there is no direct
subsidies to coal, and CIL is a profit- making entity, a comparison with three benchmark
prices helps illustrate the level of this underpricing for power generation.

The first benchmark is the price of imported coal. The weighted-average GCV of
Indian coal is about 19.6 gigajoules (GJ) per ton. The price of grade 9 coal (with a GCV

in the range of 4,600–4,900 kcal per kilogram, or 19.3–20.5 GJ per ton) is considered a representative price for Indian domestic coal. In fiscal 2011–16, the notified run-of-mine price for grade 9 thermal coal was about Rs 62.2 per GJ for the regulated sectors (including the power sector) after imposition of the Clean Environment Cess. Imported coal has a GCV of 5,200–6,500 kcal per kilogram. On the basis of its midpoint heat value of 5,850 kcal per kilogram (24.6 GJ per ton), the price of imported coal after the Clean Environment Cess ranged from $165.80 to $211.20 per GJ during fiscal 2011–16 (Figure 4.9). After adjusting for differences in heat content, the price of domestic coal for the power sector was still about half the price of imported coal in fiscal 2016.

The second benchmark is the spot market price, which has been substantially higher than the price of coal sold to power plants under fuel supply agreements—another telling indicator of the underpricing of coal for the majority of the power producers. Since November 2007, CIL has sold roughly 10 percent of its raw production each year on the spot market through an electronic auction. In addition to providing additional coal supplies to all sectors, electronic auction also works as a price discovery mechanism. Consumers in all sectors can submit bids at or above the reserve price to purchase coal on the spot market. The resulting market-driven price is substantially higher than the CIL-notified price for power plants, with the price premium ranging from 114 percent in fiscal 2012 to 33 percent in fiscal 2016.

The third benchmark is the price of coal for unregulated sectors. Under the 2018 price notification, the price of the higher grades of coal is the same for both regulated

FIGURE 4.9 Domestic coal in India is much cheaper than imported coal, even after adjusting for quality

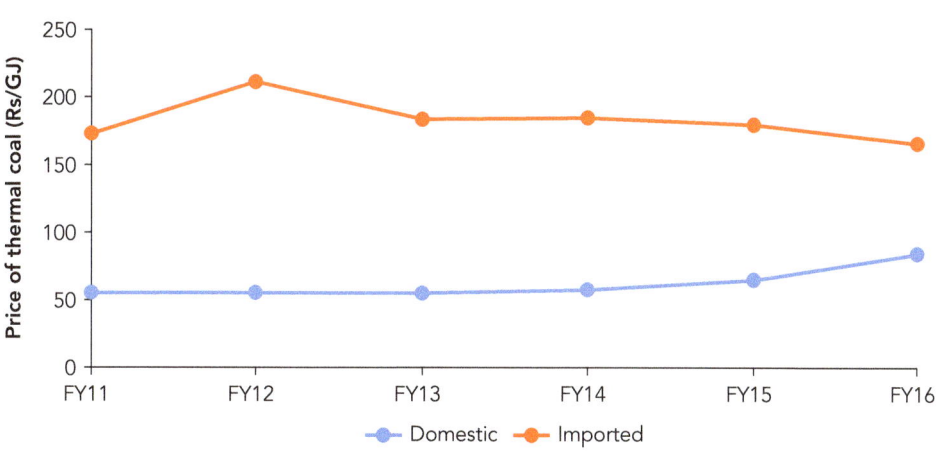

Source: Based on Coal India Limited thermal coal price notification and Ministry of Coal, India, *Provisional Coal Statistics of India* (2017).
Note: The domestic price is calculated as the weighted-average notified price of G9 coal from Western Coalfields Limited and all other Coal India Limited subsidiaries. The import price is calculated by dividing the total import value by the total import quantity. FY = fiscal year; GJ = gigajoule; Rs = Indian rupees.

FIGURE 4.10 **The price of coal is lower for power generation than for unregulated sectors in India**

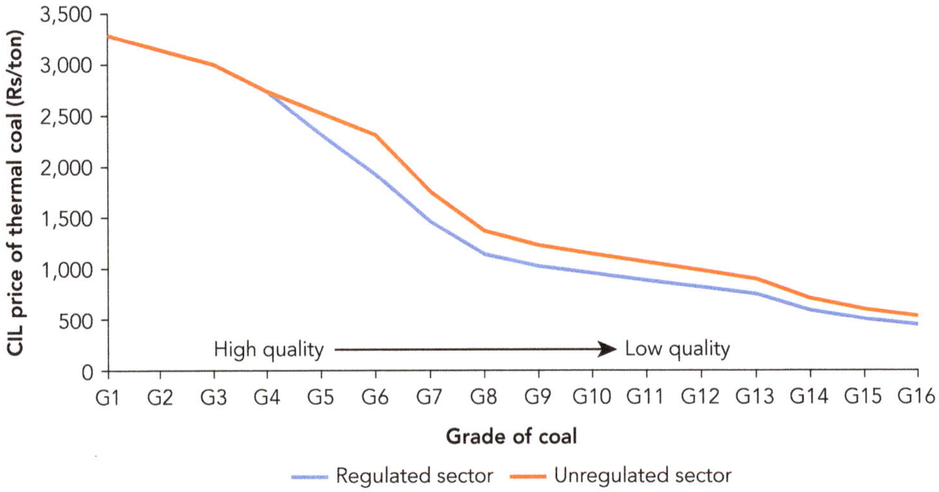

Source: Coal India Limited (CIL) price notification, January 8, 2018.
Note: Rs = Indian rupees.

and unregulated sectors (Figure 4.10). For coal with a heat value below 5,500 kcal per kilogram (a range that includes most domestic coal), the price for the regulated sectors is 17 percent lower on average than that for the unregulated sectors (CIL 2018).

Although several benchmark prices have been used to demonstrate the extent of coal underpricing in the power sector, the analysis simulates the market-clearing price in the domestic coal sector in order to estimate the deadweight loss associated with price regulation for coal in the power sector.

On the basis of the estimated supply and demand curve for coal in India (see Box 4.1), the simulation shows that, if coal-fired power plants always obtained the coal they needed from the domestic market, the market-clearing price of coal in fiscal 2016 would have been about $2.23 per GJ. This price is much higher than the administered price of about $0.84 per GJ that year but comparable to the average spot market price of $1.73 per GJ. Raising the coal price to the market-clearing price would increase the domestic coal supply by 191 million tons a year. Consumers would gain from avoided coal shortages, but they would lose from the increase in the domestic price. Producers would gain from higher sales and prices. The changes in the consumer and producer surplus from removing coal subsidies are estimated at −$0.15 billion and $2.79 billion a year, respectively. The overall welfare loss from pricing below the market-clearing price is therefore $2.64 billion (0.13 percent of GDP).

REGULATORY: COAL SHORTAGES FROM MISPRICED RAIL FREIGHT

Railways are key to transporting coal from mines to the power plants consuming it. Roughly 78 percent of India's proven coal reserves are in Odisha, Jharkhand, West Bengal, and Chhattisgarh—but only about 13 percent of coal-fired power plants are in these eastern states. Most of India's 100 largest coal-fired power plants are far from the coal mines to which they are linked through government-administered long-term fuel supply contracts. For plants not located at a pithead, this distance ranges from 150 to 1,725 kilometers, with an average distance of 647 kilometers (NTPC and Central Board of Irrigation and Power 2016).

Rail is the primary mode of transport for coal freight, but inadequate rail infrastructure has resulted in considerable bottlenecks in its delivery. During fiscal 2014, railways moved 274.3 million tons of coal across the country, almost half of the total domestic coal supplied. But 50 million tons of coal were stranded at mines because of limitations such as inadequate rail lines and a shortage of railcars (EIA 2015).

Another big problem is line congestion. India has yet to fully develop dedicated rail corridors that transport coal. To make matters worse, Indian Railways, a state-owned organization under the Ministry of Railways, gives passenger service scheduling priority over freight traffic on the country's jammed rail network, adding to the uncertainty and delay in the delivery of coal. In fiscal 2014, freight trains achieved an average speed of only about 25 kilometers an hour, or only half the speed of passenger trains (Qazi and Tahilramani 2017).

Additionally, piling up of coal at pithead also signals inefficient evacuation of coal from pithead to railhead because coal miners have an inefficient mechanism to cover transportation of coal from coal pit head to railhead which is usually 10–40 km in distance.

One of the main contributors to transport constraints is the distorted structure of rail tariffs. Historically, tariffs for passenger service in India have been kept unreasonably low at the expense of freight businesses. In 2016 passenger tickets for long-route trains were subsidized at a rate of up to 57 percent, and tickets for suburban trains were subsidized at a rate of roughly 64 percent. The cost is offset by higher freight rates, resulting in a cross-subsidy of about Rs 300 billion a year (Sharma and Sharma 2016).

Countries with efficient modern rail systems, such as Germany, have freight rates that are typically lower than passenger fares. In India, freight rates are 3.5 times higher than passenger fares. Its freight tariffs were among the highest in the world on a purchasing power parity basis, and its passenger tariffs were among the lowest (Figure 4.11). The cost of transporting coal to distant power stations is often as high as or even higher than the price of the coal at the mine (Qazi and Tahilramani 2017).

High freight tariffs have caused the rail system to steadily lose market share, particularly to roads, squeezing its revenue and its capital for much-needed investment. Freight trains use only 35 percent of the network capacity but contribute two-thirds of total revenue. Within the freight business, coal has been the largest source of revenue; in fiscal 2016 coal alone accounted for about 50 percent of the freight volume on Indian

FIGURE 4.11 India's freight tariffs are among the highest in the world—and its passenger tariffs among the lowest

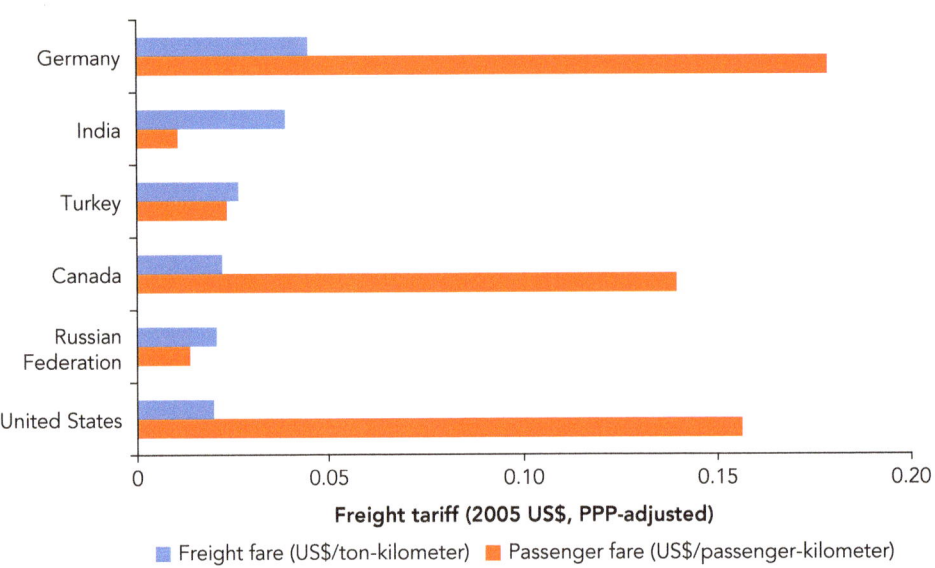

Source: Based on UIC (2012) and OECD (2013).
Note: Data are for 2010. PPP = purchasing power parity.

Railways and half the revenue-earning freight traffic (Ministry of Railways, India 2017). Lower revenue from coal freight would inevitably undermine the financial standing of Indian Railways.

How much do rail transport constraints affect coal supply for power generation in India? To address this question, the analysis explores the correlation between the distance coal travels from the mines and the shortages experienced at power stations during 2008 and 2016. It uses data from the fuel management division of India's Central Electricity Authority, which monitors the daily coal stock of India's 100 largest coal power plants. Additional data are from the Central Electricity Authority's Operation Performance Monitoring Division, which sets targets for and monitors the daily electricity output of power plants across the country, and from the National Thermal Power Corporation and Central Board of Irrigation and Power on coal linkages between coal mines and power plants.

A *coal shortage* is defined as the difference between the normative coal stock and the actual coal stock required for a power plant to operate without disruption. An *electricity shortage* is defined as the difference between a power plant's targeted output and its actual output. To smooth out daily fluctuations in coal stock and power generation caused by random noise, shortages are measured as monthly average shortages by plant.

A power plant may receive coal from more than one mine. Distance is measured as the average number of kilometers from the source of coal to a power plant, weighted by tons of coal delivered, using coal linkage data. Of the 100 power plants, 99 use rail as the mode of transport. Together, these power plants have a combined capacity of 154 GW, representing 83 percent of installed coal generation capacity in India in 2016.

Many factors could influence the magnitude of coal shortages, such as the size and age of a power station, the productivity levels of linked coal mines, the quality of coal, and seasonal and yearly effects. Indeed, the analysis finds that larger power plants and plants receiving lower-quality coal experience the worst coal shortages because they require larger normative coal stocks. There are also systematic differences between coal companies or subsidiaries. Plants linked to SCCL mines experienced the smallest coal shortages overall, whereas those linked to Mahanadi Coalfields Limited, a subsidiary of CIL, experienced the largest ones. Coal shortages peak during the monsoon season and worsened during 2008 and 2012.

To isolate the effects of distance on shortages, the analysis removes the effects of all these confounding factors. The remaining variation in coal shortages can be attributed to differences in distance. Figure 4.12 illustrates the strong positive correlation between distance and coal and power shortages. The shaded band measures the precision (or imprecision) of this estimated linear relationship.

Regression analysis shows that every 1 percent increase in distance increases a plant's average monthly coal shortage by 14 percent, all else being equal. Through its effect on coal supply, greater distance also has a negative impact on electricity generation: every 1 percent increase in distance reduces a plant's utilization rate by 3 percentage points and increases its electricity shortage, defined as the difference between targeted and actual output, by 10 percent, all else being equal.

Eliminating cross-subsidies from freight users to rail passengers would generate at least two types of benefits. First, it would lower shipping costs and the end-user price of coal for power generation.[1] On the basis of average distance traveled and the estimated cross-subsidy of Rs 0.31 per net-ton kilometer, this reduction in unit coal supply cost is estimated at $0.14 per GJ. Second, it would increase railways' revenue, which could be invested in rail infrastructure.

To gauge the potential increase in aggregate coal supply following the removal of transportation constraints, the analysis simulates a scenario in which transportation constraints are removed so that coal and electricity shortages are no longer linked to distance to coal mines. All else being equal, doing so would allow the delivery of an additional 34 million tons of coal each year.

The cost reductions and supply increases that would result from the removal of cross-subsidies would lower the marginal supply cost and shift the supply curve to the right. Reducing the unmet demand for coal would benefit consumers, and increasing the profits of coal companies would benefit producers.

FIGURE 4.12 Distance to coal mines is correlated with worse coal shortages and lower power generation in India

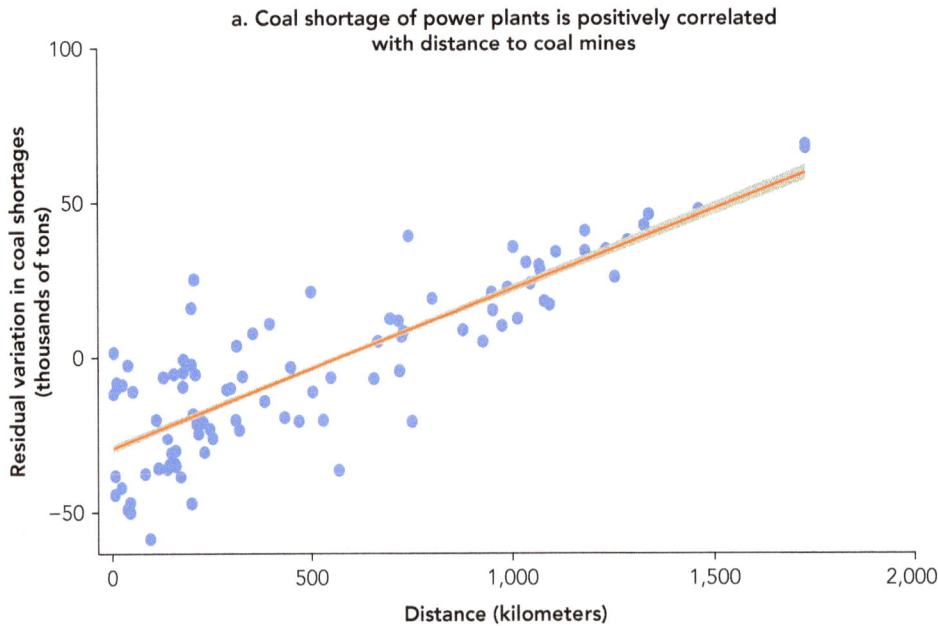

a. Coal shortage of power plants is positively correlated with distance to coal mines

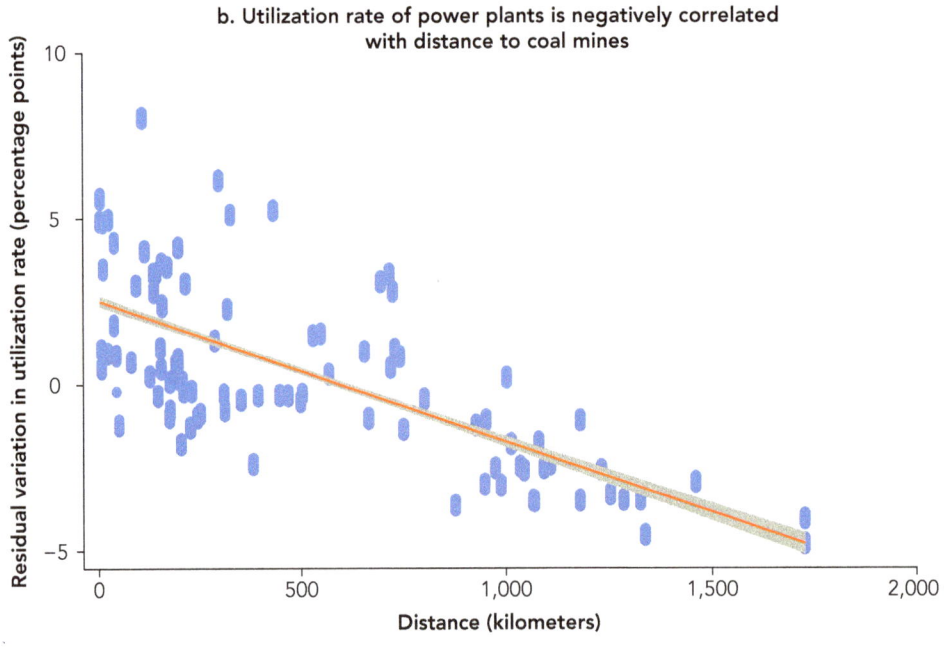

b. Utilization rate of power plants is negatively correlated with distance to coal mines

figure continues next page

FIGURE 4.12 Distance to coal mines is correlated with worse coal shortages and lower power generation in India *(continued)*

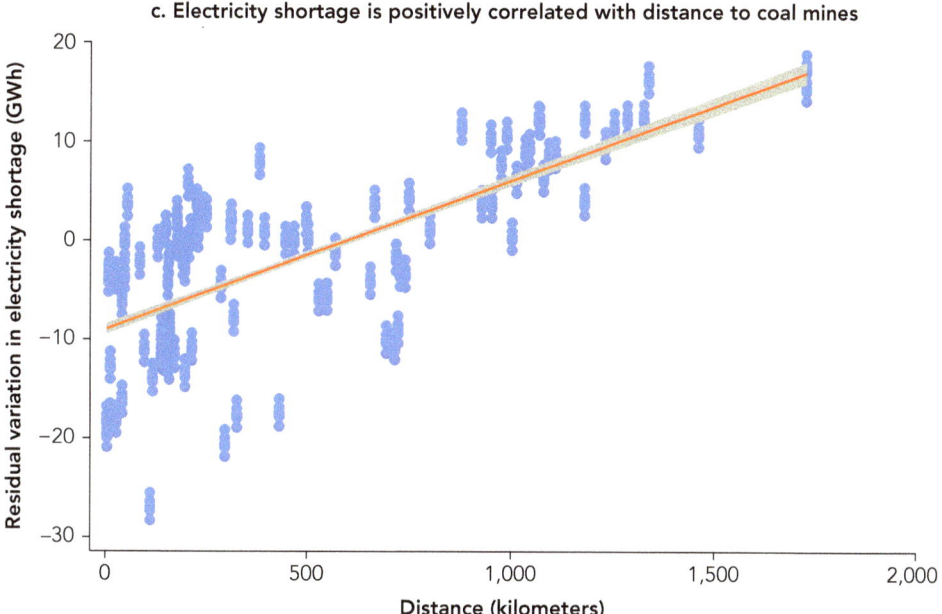

c. Electricity shortage is positively correlated with distance to coal mines

Source: Data on coal linkages: National Thermal Power Corporation and Central Board of Irrigation and Power; data on daily actual coal stock and normative required coal stock: Central Electricity Authority (2008–16); data on monthly power generation of coal plants: Central Electricity Authority (2012–16).
Note: Coal shortages are daily average shortages, defined as the normative coal stock minus the actual coal stock. Electricity shortage is defined as targeted output minus actual output. The vertical axis is the difference in residuals from regressions with and without controlling for distance between power plants and coal mines. The other independent variables in the regression are capacity; age; age squared; coal quality; and year, month, and region fixed effects. Shaded areas are 95 percent confidence intervals. GWh = gigawatt-hour.

To estimate the associated changes in the consumer and producer surplus, the analysis predicts a new supply curve based on the reduction in coal delivery cost and on the scale of the potential output increase described earlier. Using the estimated supply and demand curves for coal (see Box 4.1), the analysis suggests benefits for consumers—from a greater quantity of domestic coal—of about $901 million a year, and benefits for coal producers—from a greater quantity and a lower marginal cost—of about $529 million a year. Overall, the regulatory cost from cross-subsidization in railways is estimated at $1.4 billion (0.07 percent of GDP) a year.

SOCIAL: EMISSIONS, DISEASE, AND ACCIDENTS FROM COAL

Coal-fired power generation comes at a staggering economic cost in hidden expenses not borne by utilities. They include the health and environmental effects on mining communities and the wider society. As the most carbon-intensive source of energy, coal is a leading culprit in global climate change. Its carbon intensity (tons of carbon emitted per unit of heat content) is twice that of natural gas and 1.3 times that of diesel (IPCC 2006). India is now the world's third-largest emitter of greenhouse gases. In 2015 it emitted 2.07 billion tons of carbon dioxide, of which 72.4 percent was from coal combustion (IEA 2017a).

Coal power stations are also major sources of air pollution. The Global Burden of Disease study on India suggests that coal-burning power plants contributed to 7.6 percent of emissions of $PM_{2.5}$ annually, causing 82,900 deaths and a loss of 2.3 million years of healthy life in 2015. Air pollution has been identified as the second-highest risk factor for death in India (Global Burden of Disease MAPs Working Group 2018).

Because of the large number of people affected, the burden of disease associated with air pollution is particularly high in India. Parry and others (2014) conclude that the health damages from coal plants in India substantially exceed the climate change damages. They estimate the health damages at $190 per ton of coal—more than twice the carbon damages of $75 per ton of coal. The combined health and climate change damages from burning coal are 11 times the latest notified price of coal, which is $25 per ton after the Clean Environment Cess (Figure 4.13).

FIGURE 4.13 The market price of coal is a fraction of its social cost in India

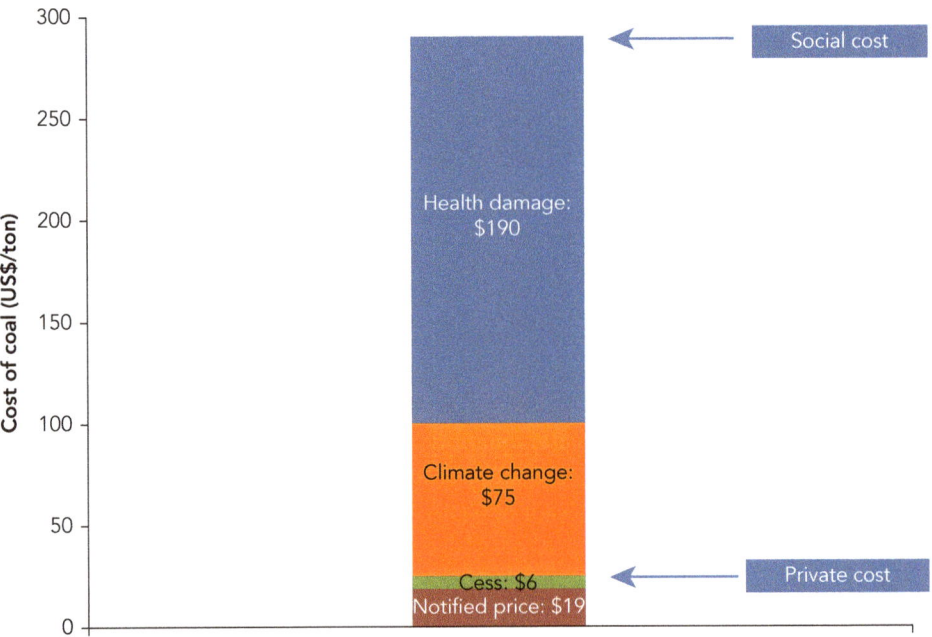

Source: Parry and others (2014).
Note: Calculations are based on the emission rate of Indian coal (1,188 tons of carbon dioxide per ton of coal) and the lower-bound estimate of the shadow price of carbon dioxide emissions ($40 per ton).

When pricing of coal fails to consider its large hidden cost, too much coal is used for power generation. To estimate the social costs of this pricing failure, the analysis calculates the potential change in welfare under full-cost pricing. The welfare calculation has three components: environmental outcomes, tax revenue, and consumer and producer surplus. Given the estimated demand responsiveness to price changes, imposing a full environmental tax on coal that reflects the cost of its negative externality (health and environmental effects) would reduce coal consumption by about 29.5 percent a year. In other words, with full-cost pricing of coal, the share of coal in the fuel mix for power generation would decline from 75 percent to 53 percent. The avoided health and environmental damages can be calculated as the reduction in consumption multiplied by the marginal damages per ton of coal, estimated at $49.9 billion a year. Imposing an environmental tax would also generate tax revenue of $116.7 billion a year. The reduction in consumption would lead to a loss in the consumer and producer surplus of $131.2 billion a year. Taken together, these results indicate that imposing the full social price would improve welfare even without considering the benefits to other countries. The net welfare gain is estimated at $35.4 billion (1.69 percent of GDP) a year in fiscal 2016.

Beyond the external costs of burning coal to produce electricity, there are also large negative externalities from extracting coal. They can be broadly classified as environmental, health, and social impacts.

The environmental impact of coal mining includes changes to the physical environment, such as landslides, soil erosion, and water pollution. Each year, more than 75 square kilometers of land in India are destroyed by coal mining (Singh 2015). The changes to the physical environment also affect land use, such as farming, tourism, hunting, and foraging.

Mining is associated with significant health and safety impacts. According to reports by the Directorate General of Mines Safety, the fatality rate in Indian coal mines declined over the last century, but it changed little in recent decades (falling only from 0.35 per thousand people employed in 1981–90 to 0.27 in 2001–10). During 1991–2009, the average rate of serious injuries was 1.75 per thousand people employed, with more than 700 such injuries on average per year, ranging from 523 in 1998 to 1,106 in 2005 (Directorate General of Mines Safety 2005, 2009). During 2001–09, the average fatality rate in India's coal mines was more than seven times that in Australia's during 2001–15 (Safe Work Australia 2018). Underground operation accounted for about 47 percent of the fatal accidents in India's coal mines in 2009 and aboveground operation for 53 percent. Coal mining also has higher mortality rates from pneumoconiosis than any other type of mining (Joyce 1998).

Coal mining also has profound effects on local communities in India. The displacement and resettlement resulting from coal mining have increased substantially since the 1970s as coal production has shifted from underground to opencast mining. Overall, mining displaced 2.55 million people in India between 1950 and 1990 (Downing 2002).

Fully quantifying the social cost of coal mining in India is beyond the scope of this report, but earlier studies using global data offer useful insights. One study estimates that coal mining accounts for 22 percent of the total social cost of coal in the United States, which also includes environmental and health costs associated with coal transport, coal combustion, waste disposal, and electricity transmission (Epstein and others 2011). A World Bank report estimates that health damages in mining communities worldwide range from $10 to $21 per gigawatt-hour of electricity produced (Grausz 2011).

Core

Shortly after independence in 1947, India began adopting legislative measures aimed at developing its core electricity sector. It created state-level, vertically integrated state electricity boards in 1948, introduced IPPs in 1991, and established CERC as an independent regulatory body in 1998. In addition, since 1996 some states have restructured their electricity industry by unbundling their state electricity board into separate generation, transmission, and distribution entities. All these states adopted a single-buyer model in which the transmission and bulk supply entity buys all electricity produced by the generators and sells electricity to the distributor.

Despite reforms, state power utilities have struggled to improve their performance. During fiscal 2001, their losses mounted to Rs 250 billion (1.5 percent of GDP). It had become clear that a statutory overhaul of the power sector was needed—one that would consolidate and replace provisions in different laws to create a single, market-oriented framework.

Two years later, India adopted the Electricity Act of 2003, which introduced substantial changes in the industry. The law mandates the unbundling of state electricity boards and the creation of independent regulatory commissions at the state level. It promotes competition through open access for transmission and distribution, enabling a switch from a single-buyer market to a system with multiple buyers and sellers. The law also proposes a multiyear approach to determining tariffs. They fluctuate only within a certain band in order to reduce regulatory uncertainty and increase incentives for power producers to reduce controllable costs. The law also calls for progressively reducing cross-subsidies and moving toward pricing based on the actual cost of supply.

How has India's power sector evolved since enactment of the 2003 law? A recent World Bank study reviewing the sector's performance finds that, despite much progress in implementing the law, many challenges remain (Pargal and Banerjee 2014). They include inefficient generation, high losses in transmission and distribution, widespread subsidies, and, most notably, sharp deterioration in the finances of state utilities.

INSTITUTIONAL: INEFFICIENT STATE GOVERNMENT–OWNED POWER PLANTS

After the introduction of the Electricity Act of 2003, there has been a dramatic increase in private sector investment. During 2012–17, the private sector contributed to 54 percent of the incremental generating capacity. However, electricity generation in India remains dominated by plants owned by the central and state governments (Figure 4.14).

FIGURE 4.14 Public power plants still dominate electricity generation in India

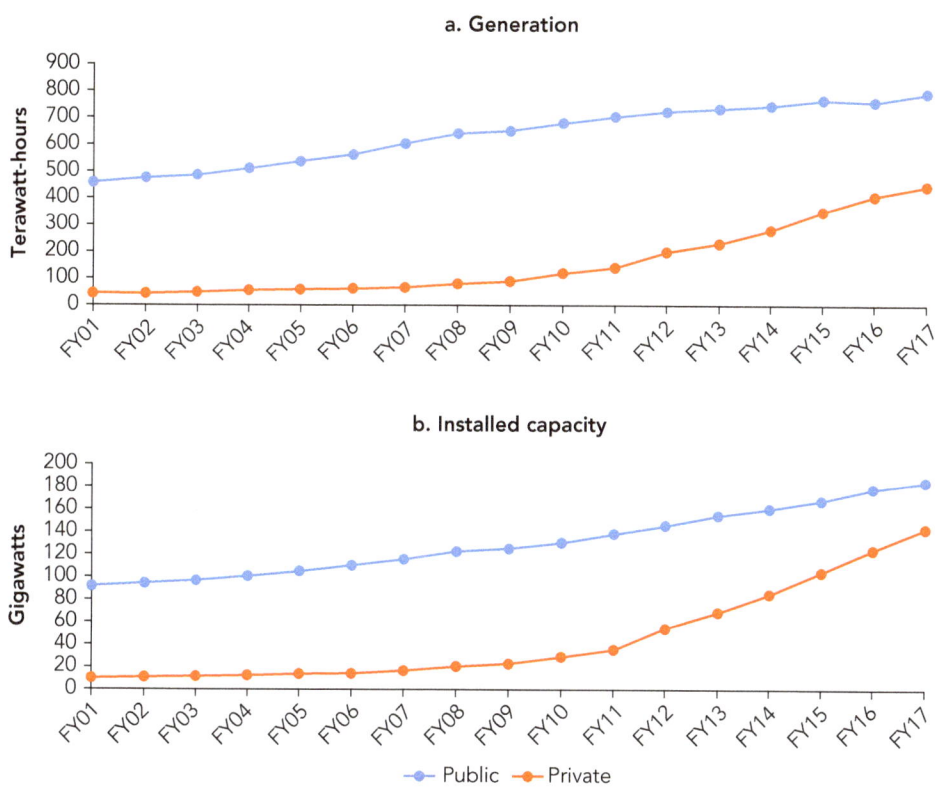

Source: Central Electricity Authority (2017a).
Note: FY = fiscal year.

In fiscal 2017, public plants represented 63 percent of total generation and 56 percent of capacity, including 61 percent from thermal power plants.

Under the guidance of the 2003 law, a competitive wholesale market has developed in which power producers can sell electricity to the highest bidder, and large customers (end-users with requirements above 1 megawatt, MW) can purchase power from the lowest-cost source. But the scope of competition is limited. In fiscal 2017, more than 90 percent of electricity was sold through long-term power purchase agreements. The rest was sold through bilateral transactions, including only 4 percent through the competitive wholesale market, the Indian Energy Exchange, and Power Exchange India (Figure 4.15).

The lack of competition in electricity supply may be related to several barriers to entry to the market. First, state governments impose a heavy open access charge on consumers (mostly high-paying and cross-subsidizing industrial and commercial users) who choose to buy electricity from a third party rather than from state distribution utilities. In several states, this additional open access charge almost doubles the cost of electricity for consumers who buy from a power exchange (Figure 4.16).

FIGURE 4.15 Most electricity in India is sold through long-term contracts

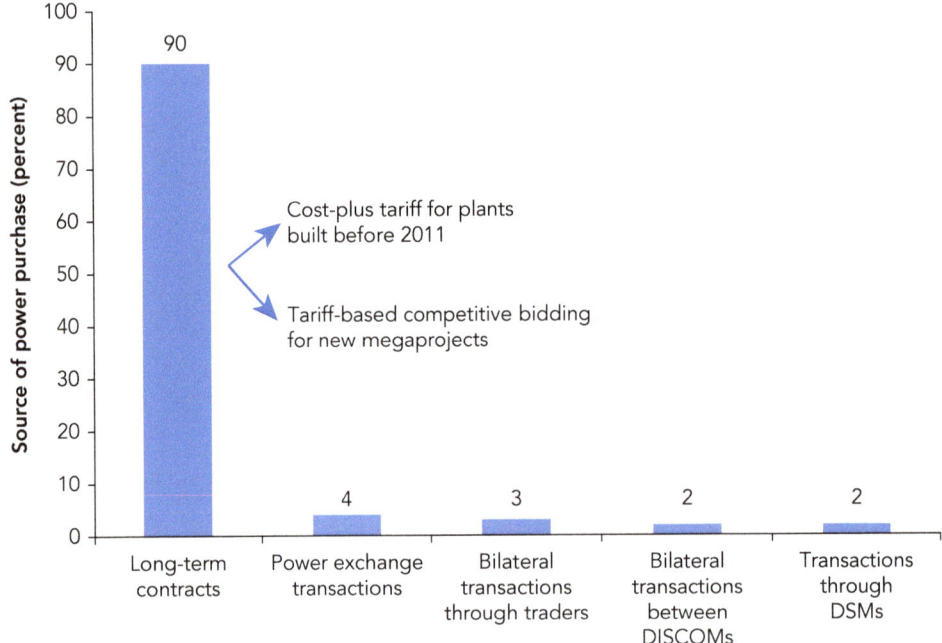

Source: CERC (2017).
Note: DISCOM = Distribution Company; DSM = Deviation Settlement Mechanism.

Second, transmission congestion has prevented power trading across states, particularly in the northern and southern regions of the country. Third, low tariffs and the distorted allocation of coal may have prevented private generators from entering the market.

When the market plays a limited role in determining price and rewarding efficiency, power plants, especially publicly owned ones, face less pressure to control costs. Analysis based on plant-level data from fiscal 2000 to fiscal 2012 for 104 coal power stations (stations with capacity of more than 25 MW, monitored by the Central Electricity Authority) reveals a large efficiency gap between public and private plants.

Two indicators can be used to evaluate the operating efficiency of a thermal power plant. The first is the load factor: the actual energy a plant generates as a percentage of the maximum possible energy it can generate in view of its nameplate capacity. This indicator measures whether a plant is maximizing output given its inputs. The second indicator is coal intensity: the consumption of coal per unit of electricity produced, expressed in kilograms per kilowatt-hour (kWh). This indicator measures whether a plant is minimizing the cost of fuel for a given level of output. Even a plant producing the maximum electricity possible would not be considered efficient if it could produce the same output with a lower level of inputs.

FIGURE 4.16 High open access charges by states deter industrial consumers from entering the wholesale market in India

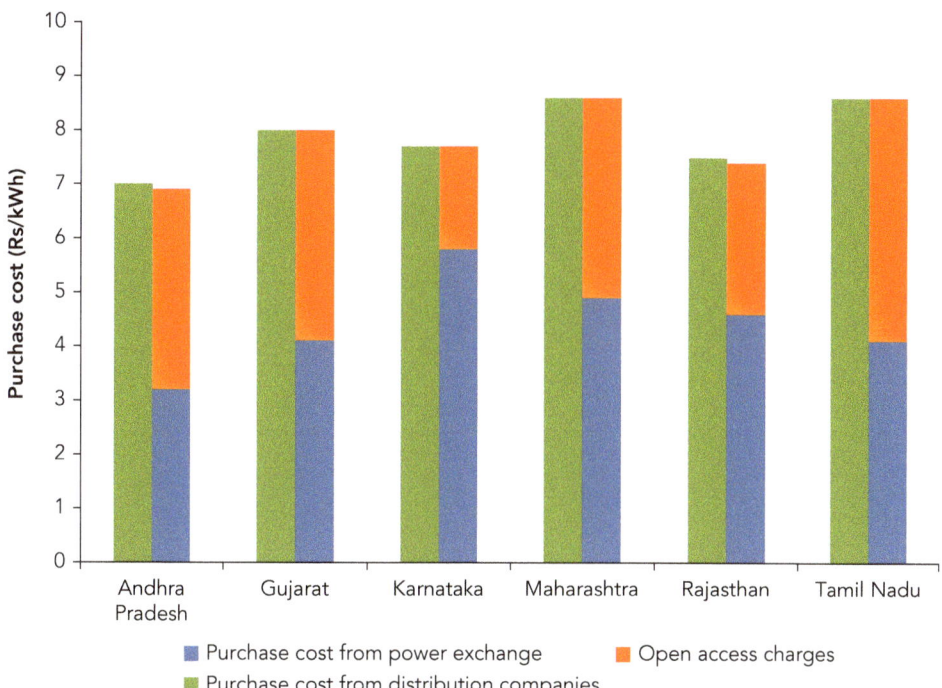

Source: Latest open access (cross-subsidy) surcharge order published on the websites of various state distribution utilities.
Note: kWh = kilowatt-hour; Rs = Indian rupees.

Cost efficiency could also be measured by the unit cost of labor and material inputs, and total factor productivity could be used in addition to partial productivity measures to assess the performance of a power plant. However, the plant-level data reported by the Central Electricity Authority do not include information on labor and material inputs, so fuel efficiency is the only indicator used to measure the cost efficiency of a power plant in India.

A simple mean comparison reveals that state-owned power plants in India are the least efficient by both output and cost measures: they run less often, and when they do operate they use more coal for a certain level of output. Centrally owned plants are more efficient than state-owned plants, but they still slightly lag behind IPPs (Figure 4.17).

Plants of different ownership types also vary significantly on a range of physical and technological characteristics. To evaluate whether factors other than ownership type drive the differences in fuel efficiency, an econometric analysis controls for variables such as the age, size, and design heat rate of power plants; the quality of coal used by the plants; and yearly shocks.

FIGURE 4.17 State-owned power plants in India are less efficient than private plants

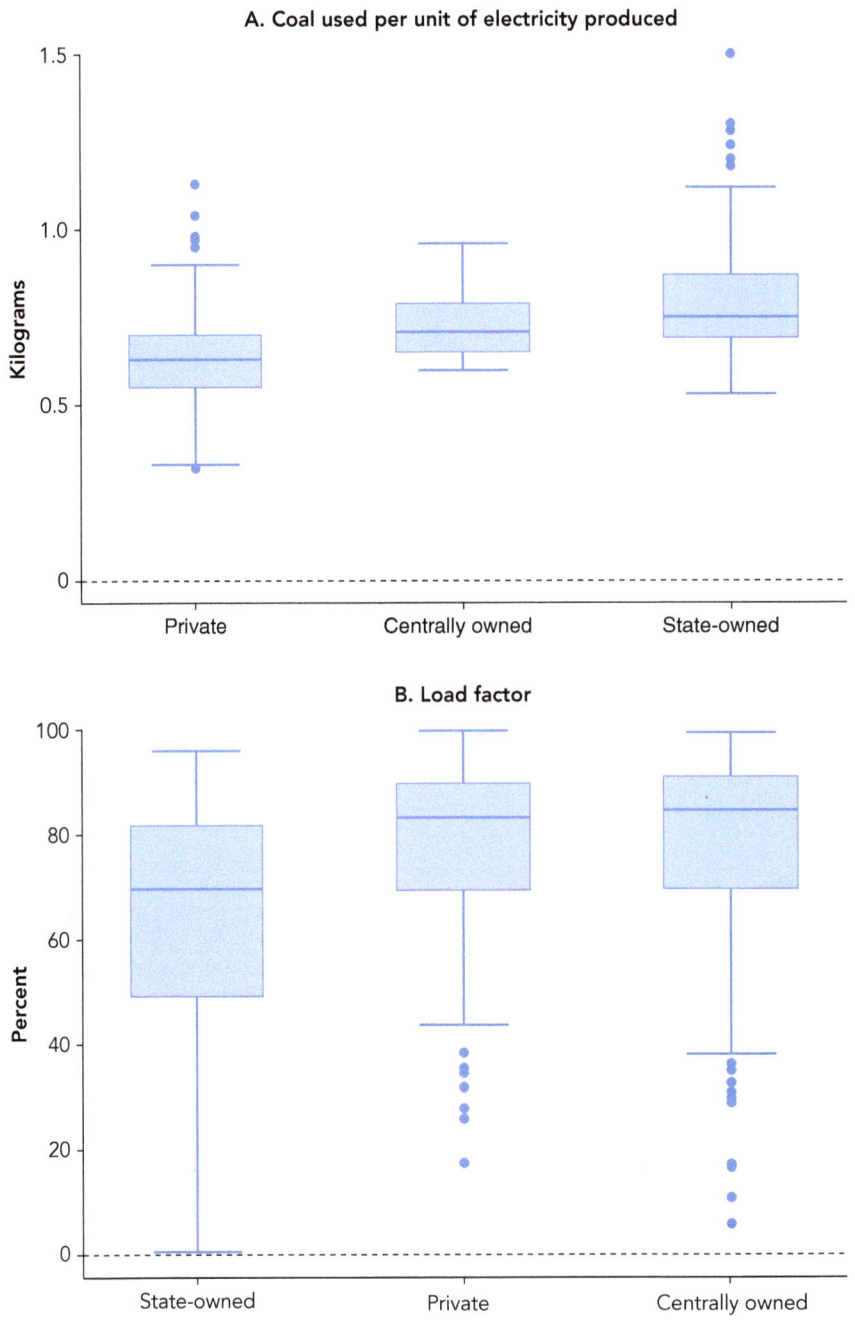

Source: Based on plant-level data, Central Electricity Authority (2000–12).
Note: Coal intensity is the ratio of coal input to electricity output. Load factor is the actual output a plant generates as a percentage of the maximum possible energy it can generate given its nameplate capacity. The graphs compare the median values of plants' coal intensity and load factors without controlling for other confounding factors.

In addition to these exogenous metrics, two other factors can affect a power plant's fuel efficiency: how often the plant is dispatched and how long it is operated below capacity. If dispatch is determined in part by ownership rather than cost, failing to control for variations in dispatch may lead to biased estimation of the level of efficiency gap associated with public ownership. The load factor can be used as a proxy for how often a plant is called to provide power, but it cannot be directly included as an explanatory variable for fuel efficiency because it is determined at least in part by a plant's fuel efficiency at the same time. To address this simultaneity concern, the analysis follows Fabrizio, Rose, and Wolfram (2007) by using statewide electricity sales as a source of exogenous variation in plants' load factor. A plant is more likely to be running when there is a statewide demand surge regardless of its ownership type, but statewide demand is unlikely to be correlated with an individual plant's coal efficiency.

Figure 4.18 presents two groups of estimated coefficients for determinants of coal efficiency. The orange bars correspond to regression analysis when controlling for dispatch through the control of statewide electricity output; the blue bars correspond to regression when not controlling for dispatch.

The results reveal that private plants, on average, used roughly 16 percent less coal per unit of electricity produced than did state-owned plants during 2000–12, all else

FIGURE 4.18 State-owned power plants in India are less efficient than private plants even after controlling for their characteristics

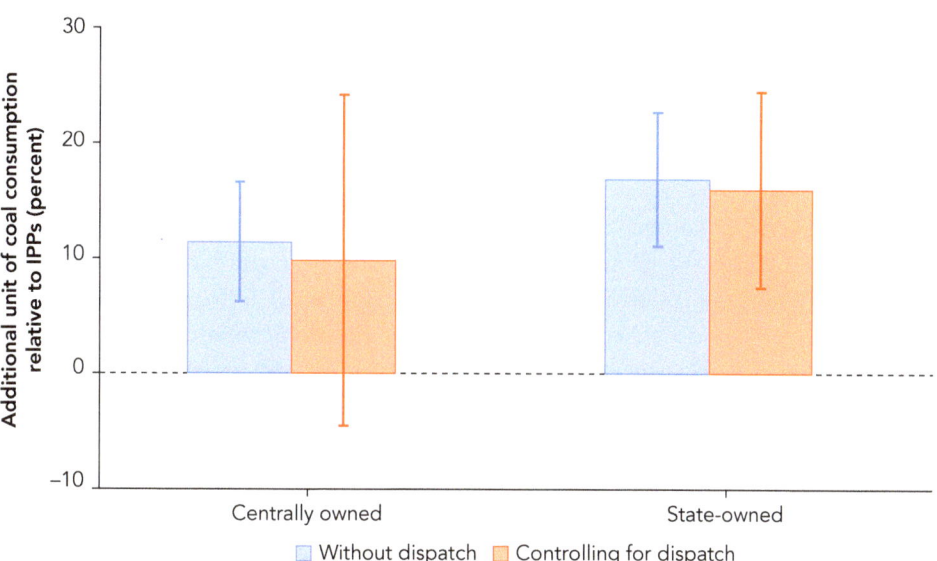

Source: Based on plant-level data, Central Electricity Authority (2000–12).
Note: Orange bars show estimated coefficients from regression analysis controlling for dispatch through the control of statewide electricity output. Blue bars show estimated coefficients from regression analysis not controlling for dispatch. Bars denote point estimates and lines denote 95 percent confidence interval. IPP = independent power producer.

being equal (including dispatch). Centrally owned plants were less coal-efficient than private plants, but the difference becomes indistinguishable from zero when controlling for dispatch. The other variables are also important in explaining differences in plant performance, and, as expected, their significance does not depend on plants' dispatch. For example, both better technology (lower design heat rate) and better quality of coal improve coal efficiency.

These results confirm that there is a large efficiency gap between state-owned and private power plants. Because the analysis has controlled for all factors related to plants' technological and operational characteristics, this efficiency gap is likely to be correlated with other innate differences between plants. For example, anecdotal evidence suggests that some private plants have signed power purchase agreements that have allowed them to be dispatched only at the optimal load factor for maximum fuel efficiency. However, because power purchase agreements are mostly confidential, the extent of such preferential treatment is unknown.

To the extent that the efficiency gap is also related to the difference in managerial behavior between plants, improving the quality of management of state-owned power plants to match that of private plants would reduce coal shortages and mitigate their effects on the power supply. To estimate the size of such a potential impact, a simulation focuses on the northern region of the Indian power system—a system on which detailed data on power shortages and plant dispatch are reported daily by the Northern Regional Load Dispatch Center. Because demand and shortages vary throughout the day, an hourly resolution of the supply and demand profile is required to pin down how much and when a production increase can be used to alleviate shortages. The efficiency improvement in each state-owned power plant (a 16 percent increase in fuel efficiency) is applied equally in each hour of the day, but only up to a plant's installed capacity. Depending on whether shortages are present during an hour, the additional output is used first to reduce unserved energy demand and then to offset output from more expensive plants. The simulation analysis also takes into account a 17 percent network loss, which would offset some of the gains.

The results show that reduction of the coal intensity of state-owned power plants by 16 percent to match the managerial performance of IPPs would have reduced power shortages in India's northern region by 46 percent in fiscal 2015 (Figure 4.19). The efficiency gain would also yield cost savings, reducing the unit generation cost by 4.2 percent a year and the total generation cost by $70 million. The estimated total cost savings are relatively modest because of the assumption that most of the gain in efficiency would be used to make up for shortages rather than to replace inefficient plants. Extrapolating the results to all of India (on the basis of a regionwide generation profile) suggests that improving the efficiency of state-owned power plants could reduce power shortages by 50 percent, or by 21 terawatt-hours (TWh) using fiscal 2015 data.

The increase in fuel efficiency would shift the aggregate supply curve of electricity to the right. The analysis simulates this new supply curve based on the scale of the potential output increase described earlier and the estimated price elasticity of electricity

FIGURE 4.19 Improving the fuel efficiency of state-owned power plants would significantly reduce power shortages in India

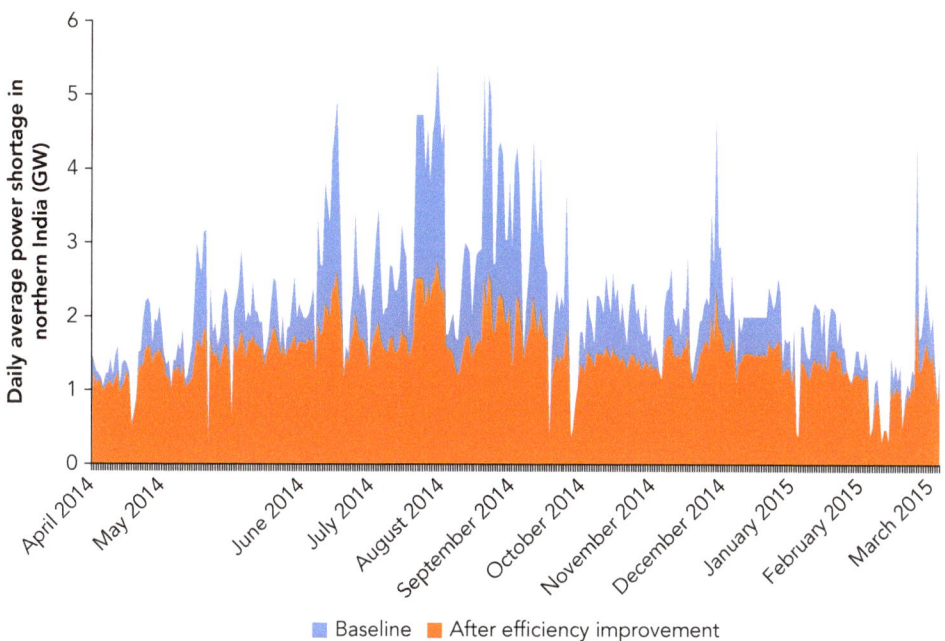

Source: Simulation based on Central Electricity Authority (2000–12) and daily reports by the Northern Regional Load Dispatch Center.
Note: GW = gigawatt.

supply and demand in India (Box 4.2). It then projects the new equilibrium of quantity and price. Results from the simulation reveal a benefit of $3.4 billion a year for both consumers and producers from the lower production costs and greater quantity of electricity. These benefits are partially offset by increases in the government's subsidy spending. The net institutional cost from the inefficiency of state-owned power plants is estimated at $2 billion (0.10 percent of GDP) a year.

It should be noted that the above analysis is based on data between fiscal 2000 and fiscal 2012, the most recent disaggregated plant-level data that are available at the time of analysis. It is possible that the efficiency gap between state-owned and IPPs has since changed. Indeed, official figures of aggregated data show that the average coal efficiency of state-owned power plants, without controlling for differences in plants' physical and technical characteristics, is roughly 12.5 percent lower than that of IPPs during 2014–17. This number is smaller than the mean average efficiency difference between state-owned and IPPs between fiscal 2000–12, which was about 17 percent.

Furthermore, a market-based energy efficiency program, Perform, Achieve and Trade (PAT) scheme was introduced in India in 2012. Under this scheme, targets were set for energy-intensive sectors to save about 6. 5–10.0 million tonnes of oil equivalent (mtoe)

BOX 4.2 Estimating the price elasticity of supply and demand for electricity in India

The analysis estimates the price elasticity of electricity supply and demand in India, using income and price as the main determinants of supply and demand. The estimation is based on annual data for electricity sales, estimated electricity shortage, the weighted-average electricity price index in real terms, and real per capita GDP over the period 1974–2013. These data are from various annual reports of the Central Electricity Authority; the website of the Ministry of Statistics and Program Implementation; *On the Working of State Power Utilities and Electricity Departments*, published by the India Planning Commission; and monthly reports of the National Load Dispatch Center, accessed through Indiastat. To check for robustness, the analysis also estimates the price elasticity of electricity demand in the residential sector using household-level panel data from the Indian Human Development Survey for 2005 and 2012.

To address the potential endogeneity of price, the contemporaneous electricity price is instrumented by the first six lags of electricity price in the estimation of the supply curve. To estimate electricity demand, the number of annual cooling degree days is used as an exogenous demand shifter, and the first four lags of electricity price are used as instrumental variables for the contemporaneous electricity price. The analysis tests for the existence of long-run cointegration and estimates an error correction model to obtain both short- and long-run elasticities (Appendix A provides details on the methodology).

The long-run supply and demand elasticities are estimated as 0.39 and –0.42, respectively, meaning that a 1 percent increase in the price of electricity would increase the supply of electricity by 0.39 percent and reduce the demand for electricity by 0.42 percent. These estimates are in the range of elasticities suggested by the empirical literature on electricity demand in developing countries (Zhang 2015). They are also in line with estimates derived from disaggregated data. On the basis of a regression analysis that controls for household demographic and housing characteristics, and an appliance portfolio, the analysis estimates the long-run price elasticity of residential demand for electricity at –0.48. Using the National Sample Survey for 1993/94, Filippini and Pachauri (2004) find that the short-run price elasticity for urban households is –0.42 during winter, –0.51 during the monsoon months, and –0.29 during summer.

of energy, out of which 3.1 mtoe of energy saving is expected from the thermal power sector alone (CII, Confederation of Indian Industries 2011). As of today, 208 thermal power plants have participated in the scheme, including 87 state-owned power plants. These plants have taken energy efficiency measures to meet performance targets. As a result, efficiency ranking among plants of different ownership types may also have changed. Further analysis on the normalized efficiency gap between state-owned power plants and IPPs is warranted when plant-level data for recent years become available.

INSTITUTIONAL: UNDERINVESTMENT IN TRANSMISSION

Lack of investment in transmission and distribution infrastructure has resulted in congestion of the network in India, impeding the evacuation of power and the development of a competitive power market. India's transmission network consists of five regional

TABLE 4.1 Transmission congestion hinders the evacuation of power in India

Fiscal year	Unconstrained power transaction (TWh)	Actual power transaction (TWh)	Unsold electricity because of congestion (TWh)	Share of unsold electricity in unconstrained power transaction (percent)
2010	8.10	7.09	1.01	12.5
2011	14.26	13.54	0.72	5.0
2012	17.08	14.83	2.25	13.2
2013	27.67	23.02	4.65	16.8
2014	35.62	30.03	5.59	15.7
2015	31.61	28.46	3.14	9.9
2016	36.36	34.20	2.16	5.9
2017	41.60	40.08	1.52	3.7

Source: CERC (2017).
Note: TWh = terawatt-hour.

grids—northern, eastern, southern, western, and northeastern—that were synchronously connected by December 2013 and operate at one frequency. Table 4.1 summarizes the annual congestion in the power exchanges from fiscal 2010 to fiscal 2017. Although power curtailment caused by transmission constraints has gradually declined since fiscal 2013, a significant amount of electricity continues to be lost to congestion in the electric network. If there had been no transmission congestion in fiscal 2017, almost 4 percent more electricity could have been transmitted in the power exchange than was actually cleared that year. Such bottlenecks occur mostly in the northern and southern regions.

Similar to the previous exercise, the analysis simulates how much consumer and producer surplus would increase if more transmission capacity were built to remove congestion in the network. It assumes that, when there is no transmission constraint, the wholesale market is fully competitive. It also assumes that transmission capacity expansion would increase demand by 2.16 TWh in fiscal 2016, equivalent to the same amount of electricity lost in the wholesale exchange from congestion that year. The simulation reveals that the net increase in market surplus is $250 million a year.

Besides creating power shortages, transmission constraints can serve as a source of market power because they enable suppliers to raise prices in the thinner regional markets that are experiencing transmission bottlenecks. Ryan (2017) simulates bidding outcome in a competitive market. He finds that the deadweight loss from such market power is $110 million a year in India. The total cost from underinvestment in transmission network is therefore estimated at $360 million (0.02 percent of GDP) a year.

INSTITUTIONAL: HIGH LOSSES OF DISTRIBUTION UTILITIES

Implementation of the Electricity Act of 2003 in the distribution sector remains uneven across states. By 2013, 28 state electricity regulatory commissions were operating across the country (with Manipur and Mizoram sharing one), setting tariffs, establishing performance standards, and protecting consumer rights. Eighteen states had completed the unbundling process to varying degrees, with 11 ending up with multiple distribution companies, 6 having only one distribution company, and 3 having separated only transmission operations while generation and distribution still functioned as one utility. Utilities in the other 10 states continue to operate as single entities. The National Capital Territory of Delhi also unbundled to multiple distribution companies.

Privatization remains limited in the distribution or retail market compared with the generation sector. The National Capital Territory of Delhi and the state of Odisha privatized distribution in the early 2000s. Odisha's distribution utilities were returned to government control in 2015 following a long period of unsatisfactory performance. Seven other states established private distribution companies, and 42 distribution utilities have been corporatized. The other 10 distribution utilities still operate as state power departments.

Regardless of the status of unbundling or privatization, all distribution companies act as regional monopolies. Although all states have instituted open access, less than 15 percent of eligible consumers purchased electricity from IPPs rather than from distribution companies in 2016.

An uncompetitive market, combined with dominant government ownership, creates concerns about accountability, incentives, and efficiency. Moreover, although many states have unbundled and corporatized their utilities, state governments continue to interfere in utility operations, undermining the principle of commercial operation. They also add to the financial difficulties of utilities by pressuring them to keep tariffs low and to buy expensive power to cover short-term deficits during elections (Pargal and Banerjee 2014; World Bank 2013). These problems erode the functional independence of the unbundled entities, making it difficult to hold them accountable for their performance. Meanwhile, repeated bailouts from the government create a soft budget constraint, further weakening incentives for improving efficiency.

What is the level of operational inefficiency in distribution? One way to measure it is by the share of electricity losses incurred in supplying electricity—that is, transmission and distribution losses. Although these losses have gradually declined since fiscal 2001, they remained as high as 22 percent in fiscal 2016, among the highest in the world (Figure 4.20). Poor infrastructure, faulty metering, and outdated equipment all contributed to high network losses. In addition, electricity theft is rampant in India, and so a large share of these losses is likely to be commercial losses.

Although no official estimate of the relative size of commercial losses is available, the Indian Human Development Survey offers some hints. Because the survey asks households what mode of payment they use for electricity connections and how much they pay, it allows an estimate of the percentage of nonstandard connections associated with households that reported having access to electricity but either did not receive bills and did not make a payment or paid neighbors. These data suggest that about 15 percent of connections were nonstandard in 2005; this share fell to 9 percent in 2012.

High electricity losses have contributed to the deteriorating financial situation of distribution companies. Between fiscal 2010 and 2016, losses booked by utilities after

FIGURE 4.20 Transmission and distribution losses are high in India

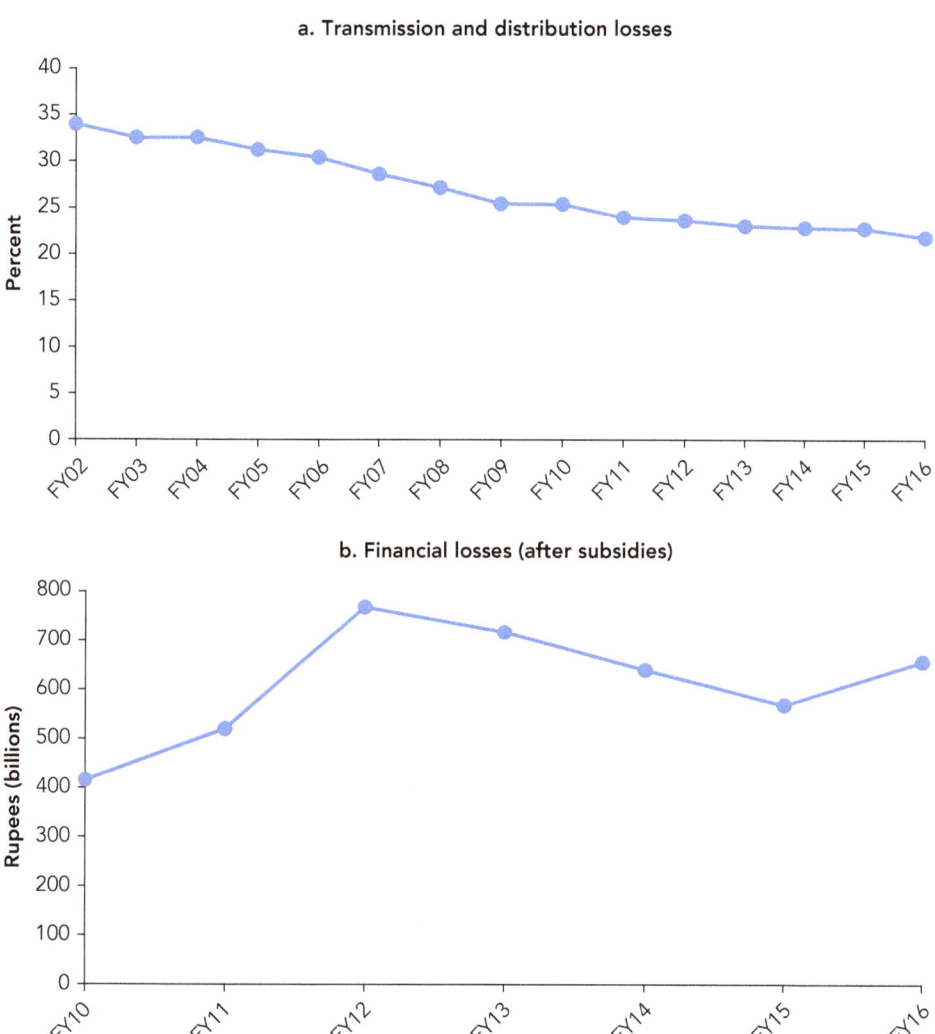

Source: Based on data from Central Electricity Authority (2017b) and Power Finance Corporation (2015, 2017).
Note: FY = fiscal year.

receiving subsidy payments increased by roughly 58 percent. Pargal and Banerjee (2014) decompose the financial losses of distribution utilities into three parts: distribution losses (including pilferage), low tariffs, and collection losses. They find that distribution losses were the largest contributor to the total losses of distribution companies in 2003–11. Tariffs that were prescribed below the full cost of operation were another important contributor (see the section on Underpriced Electricity).

As described, many factors affect transmission and distribution losses, including managerial performance (possibly influenced by ownership and unbundling status), size of the workforce, state of the distribution network, adoption of modern technology, and density of the population served. To gauge how much electricity was lost to poor managerial performance, the analysis estimates a production frontier for distribution in India and calculates the distance to the production frontier for each utility.

Distance to the frontier (the best feasible performance), also called *technical effi-ciency*, indicates how much more electricity could have been sold had there been no inefficiency. It is assumed to be a function of the operational type of utilities. The pro-duction frontier is estimated using a stochastic frontier approach and data reported by the Power Finance Corporation for 58 distribution utilities from fiscal 2012 to 2016, the most recent data available at the time of analysis (see Appendix B for methodological details). These data include total electricity sold (as output), capital and labor inputs, and total amount of electricity feasible for sale. The analysis also controls for the geographic location of distribution companies, their operational type (department, corporatized, or privatized), whether they are bundled or unbundled, and a time trend.

Unsurprisingly, the results show that most distribution companies fall well short of the production frontier. The average technical efficiency score is 0.85, meaning that, for a randomly selected distribution utility, the actual sale of power is 85 percent of the maximum feasible sale (Figure 4.21). Technical efficiency varies significantly between utilities, even after controlling for differences in capital, labor, and location. The differ-ences in efficiency can therefore be potentially ascribed to differences in the managerial efforts of utilities and in other institutional characteristics such as autonomy and trans-parency (Pargal and Mayer 2015).

How costly is the operational inefficiency in distribution associated with institu-tional shortcomings? To estimate this cost, the analysis considers a counterfactual in which all utilities improve their technical efficiency to match the best performance observed in the sample (Punjab State Power Corporation Limited achieved the high-est technical efficiency score in fiscal 2016). This scenario would result in the sale of an additional 40 TWh of electricity a year. Removing operational inefficiency would shift the supply curve to the right. The analysis predicts the new equilibrium under efficient operation on the basis of the potential increase in output. The corresponding change in consumer and producer surplus net the increase in spending on subsidies is estimated at $2.0 billion (0.10 percent of GDP) a year.

One caveat is that theft is responsible for a large share of electricity losses in the network. Thus, even though reducing losses through better billing and payment enforcement could help improve both the financial standing of distribution utilities and the overall health of the power sector, the actual increase in power supply may not be correlated linearly with the reduction in transmission and distribution losses. However, even the best-performing utility in the sample may not be efficient when compared with international best practice. If transmission and distribution losses were reduced to the international standard of 10 percent, the potential gain would be much larger.

It is worth mentioning that the Government of India has launched several reforms aimed at bringing down losses occurred in electricity transmission and distribution. For example, the Restructured Accelerated Power Development and Reforms Program

FIGURE 4.21 Technical efficiency scores vary widely across distribution utilities in India

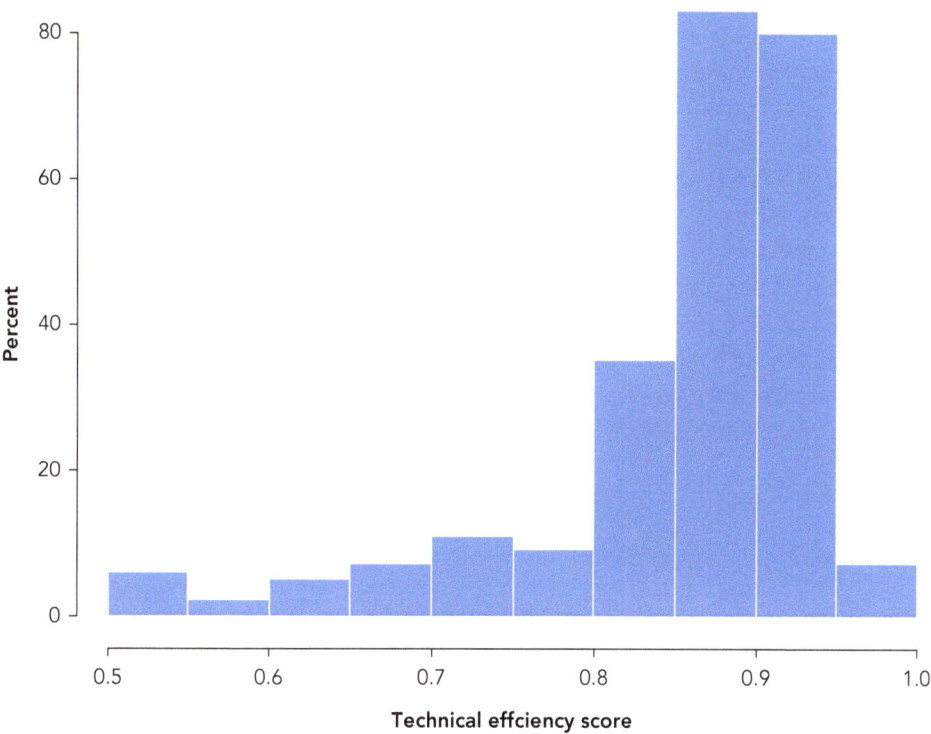

Source: Based on data from Power Finance Corporation (2012–17).
Note: The technical efficiency score is the ratio between actual sales and the maximum feasible sales of electricity.

(R-APDRP) supports the modernization and strengthening of transmission and distribution network, the adoption of information technology for data collection and monitoring, and the provision of capacity building and incentive scheme for distribution personnel to reduce average technical and commercial losses. The Integrated Power Development Scheme (IPDS) launched in 2015, among others, aims at strengthening sub-transmission network and metering, customer care services, and the completion of the ongoing works of R-APDRP. All distribution companies are eligible for financial assistance under the IPDS scheme. Successful implementation of these schemes would contribute to network loss reduction over time.

REGULATORY: UNDERPRICED ELECTRICITY

India subsidizes electricity for farmers and households, as is evident in its tariff structure. On average, a gap persists between the cost of electricity supply and the tariffs charged to these two categories of consumers (Figure 4.22). In fiscal 2016, for example, when the average cost of supply across all consumer groups was Rs 5.43 per kWh, the tariff for agricultural use was only Rs 1.71, or roughly 31 percent of the cost. The tariff for the residential sector was a bit higher, at Rs 4.08 per kWh, but still only about 77 percent of the average cost of supply. In fiscal 2016, the weighted-average tariff for all categories of consumers at Rs 4.23 per kWh was 22 percent lower than the weighted-average cost of supply (Power Finance Corporation 2017).

This gap is financed in part by cross-subsidies and in part by government budgetary support. In fiscal 2016, total fiscal spending on subsidies for electricity reached Rs 552.8 billion, 56 percent of which went to agriculture and 44 percent to residential consumers. Making matters worse, subsidies to utilities are not always paid on time. During fiscal 2016, the difference between subsidies booked and subsidies received reached Rs 24.0 billion, adding to the financial woes of distribution companies.

The fortunes of the power sector are closely linked with the performance of the distribution sector because that is where revenues are generated. Khurana and Banerjee

FIGURE 4.22 India subsidizes the electricity price paid by residential and agricultural consumers

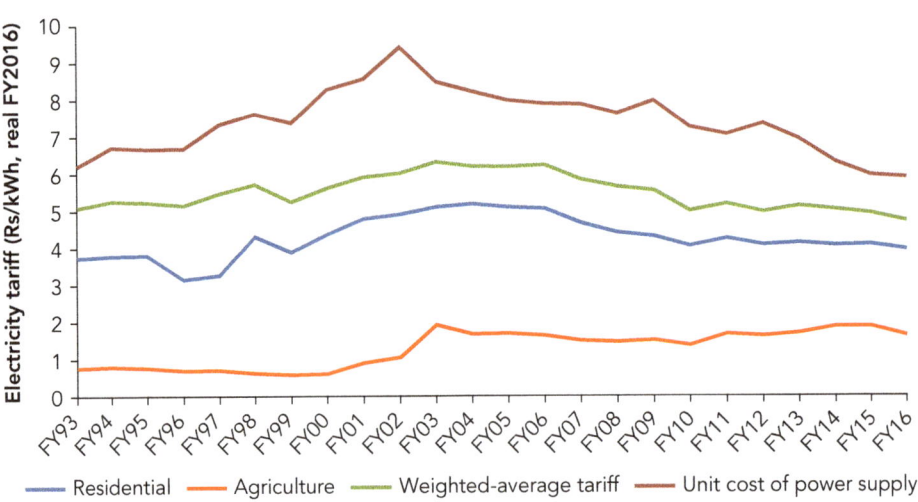

Source: Based on data from Power Finance Corporation (2004, 2005, 2016) and Indian Planning Commission Power and Energy Division (2000, 2001, 2002, 2012, 2014).
Note: Unit cost of power supply for fiscal 2015 and 2016 are based on annual plan projections. Domestic prices for fiscal 2002 and 2003 were not available and were interpolated. FY = fiscal year; kWh = kilowatt-hour; Rs = Indian rupees.

(2014) find that the underpricing of electricity drove a sharp increase in the losses booked by distribution utilities during 2003 and 2011. Indeed, in 2011 the gap between revenue and costs accounted for 14 percent of total losses reported.

Underpricing electricity threatens the reliability of supply. It not only undermines the ability of distribution companies to purchase electricity and to carry out maintenance and investment but could also create perverse incentives for utilities to underserve loss-making customers, especially in rural areas where the cost of service is high. Using data from the electricity sector in Colombia, McRae (2015) finds that subsidies deter investment in modernizing infrastructure and trap households and utilities in a nonpayment, low-quality equilibrium. Similarly, analysis using nightly satellite images from India for 2013 finds that areas adjacent to newly electrified villages subsequently experienced worse power outages (Box 4.3). One possible explanation for this finding is that, as more low-paying consumers joined the grid, a greater strain was placed on it because distribution utilities were either unable or unwilling to invest in maintaining and upgrading infrastructure to expand the power supply.

Pricing below cost-recovery levels also results in a weak price signal for energy conservation, leading to excessive consumption. But in India today, because demand for electricity is depressed by persistent power shortages, an increase in its marginal price will have two opposing effects on its consumption. On the one hand, it will reduce consumption as a result of the higher price; on the other hand, it will increase consumption because of a greater supply of electricity and a possible greater willingness to pay for it—that is, both the demand and the supply curve shift out. For example, Banerjee and others (2015) show that outages can discourage households from paying a monthly fixed fee for access to the electricity grid.

The dynamics between the electricity price and the relative positions of the supply and demand curves are complex and difficult to model. Analysis of the cost of underpricing electricity therefore focuses on estimating the static deadweight loss. The analysis assumes that consumers receive a subsidy that lowers the price of electricity by 22 percent. Producers are fully paid after the government budgetary transfer. The electricity supply falls short of demand by 2.1 percent, or 24 TWh, in fiscal 2016. On the basis of the estimated long-run supply and demand elasticities for electricity (Box 4.2), the analysis predicts the change in consumption and profits that would result from raising the electricity price to the market-clearing level. Both supply and demand would decline, and unmet demand would be eliminated. But the reduction in taxpayer spending as a result of the elimination of electricity subsidies would more than offset the decrease in consumer and producer surplus. The static gains from removing electricity subsidies are estimated at $321 million (0.02 percent of GDP) a year.

Eliminating electricity subsidies reduces the fiscal burden. It also can help break a "subsidy trap" by facilitating a shift from a low-payment, low-quality equilibrium to a high-payment, high-quality equilibrium. The change can lead to actual welfare gains that are much greater than the static welfare gains reported here.

BOX 4.3 Whose power gets cut in India?

When and where electricity is provided can have important effects on welfare and growth. But quantifying those effects is difficult because utility-level data on power outages are rarely available and not always reliable. Using big data techniques, a team from the World Bank and the University of Michigan developed a new method of detecting power outages from outer space. It involves identifying outage-prone areas through measures of excess fluctuations in light output across a long time series of nighttime satellite imagery.

To develop these measures, the team acquired and processed the complete historical archive provided by the U.S. National Oceanic and Atmospheric Administration of the nighttime satellite imagery of the suborbital Defense Meteorological Satellite Program's Operational Linescan System (DMSP-OLS) captured over South Asia every night between 1993 and 2013. These images reveal concentrations of lighting at a fine spatial resolution of 0.56 kilometers and smoothed resolution of 2.7 kilometers. The DMSP overpass time is between 7:30 p.m. and 10 p.m. Studies have shown that DMSP-OLS can reliably detect electrified villages in developing countries and that nighttime light output is a useful proxy for electricity provision (Min and others 2013; Min and Gaba 2014). The analysis is updated through 2013 because DMSP data after 2013 are considered not useful because of increased solar glare during many periods of the year.

From the raw images, the team extracted the level of nightly light output observed over each of India's 600,000 villages. It then processed these data by dropping "bad" data, including observations compromised by cloud cover and stray light, and removing background noise from non-ground-based factors such as sensor noise or atmospheric conditions. The background noise is calculated as the average brightness recorded in a sample of unpopulated and unelectrified areas in India that should not be emitting any light at night from the ground.

To identify the intensity of power outages, the team computed the mean and the standard deviation of the recalibrated brightness values for each village-year. It then regressed the standard deviation of the brightness value on its mean and additional polynomial terms. The residuals of these regressions are defined as the power supply irregularity (PSI) index. Positive values indicate higher variability in light output than expected given the average light output; negative values represent more stability in light output.

The team computed annual estimates of the PSI index for all 600,000 villages in India from 1993 to 2013. Map B4.3.1 shows the PSI values in 2013 for individual villages in several states of India. These measures are consistent with ground-based measures of power supply reliability from the Indian Human Development Survey and with feeder-level outage data from the Maharashtra State Electricity Distribution Company, which provides disaggregated data for the system average interruption frequency index for each month from 2009 to 2013.

The team also acquired comprehensive data from the Indian government listing all villages qualifying for participation in the national electrification program, Rajiv Gandhi Grameen Vidyutikaran Yojana (RGGVY), at the end of 2012. Analysis comparing power outages in 2013 against prior village electrification efforts under RGGVY during 2005–12 finds that outages were higher in districts that had large numbers of newly electrified villages. In addition, the irregularity of power supply increased in previously electrified villages as the rate of neighboring RGGVY participation increased (Min 2016). These findings suggest that

box continues next page

BOX 4.3 **Whose power gets cut in India?** *(continued)*

the extension of the electricity grid in 2005–12 may have led to an increase in the intensity of power outages in 2013.

MAP B4.3.1 **Power supply irregularity index of villages in India in 2013**

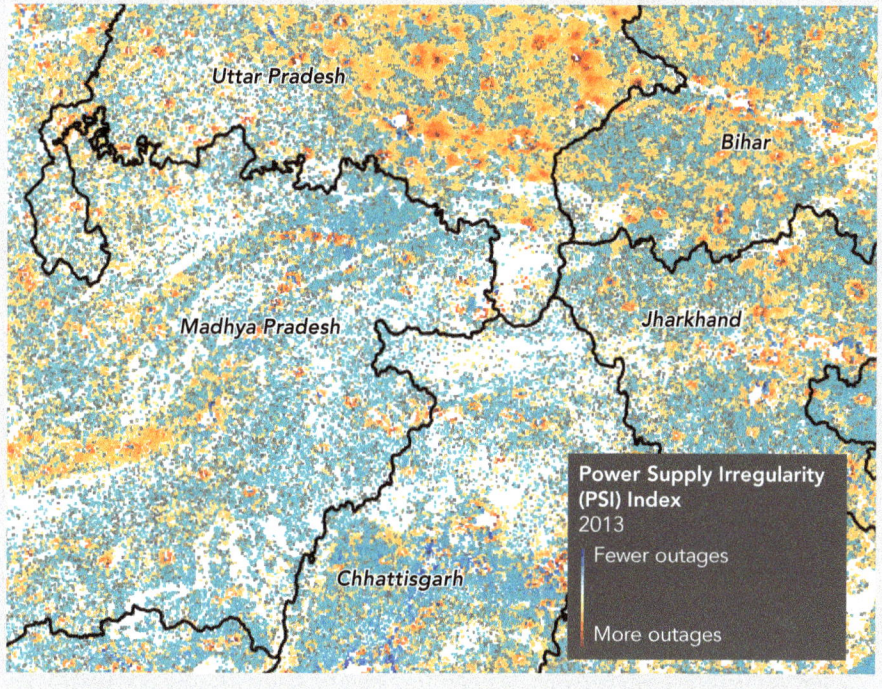

Source: Min, O'Keeffe, and Zhang (2017).

SOCIAL: GROUNDWATER DEPLETION FROM CHEAP ELECTRICITY

India heavily subsidizes electricity tariffs for farmers. To promote agricultural production, most state governments have since the 1970s adopted flat-rate electricity tariffs for farmers and provided unmetered power for pumping irrigation water. Under flat-rate tariffs, farmers pay a fixed monthly fee for electricity based on the capacity of their water pumps. Although the implicit rates suggested by the fixed monthly charges vary widely across states, all fall well below the average supply cost of $0.07 per kWh (Figure 4.23).[2]

In fiscal 2016, agricultural consumers accounted for about 22 percent of total electricity sales but only 8 percent of revenue. Unsurprisingly, for distribution utilities a higher volume of sales to agriculture is associated with a higher supply cost and lower revenue. Using data for 59 distribution utilities for 2007–13, the analysis finds that, when the share of agricultural sales goes up by 10 percent, revenue without subsidies goes down

FIGURE 4.23 Most states in India highly subsidize the price of electricity for farmers

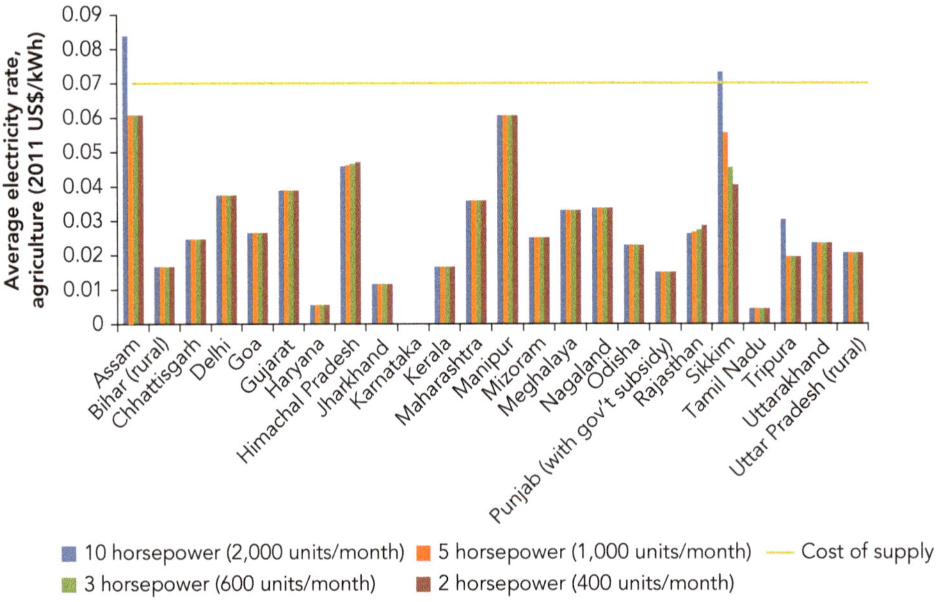

Source: Ministry of Power, India, accessed through Indiastat.
Note: Bihar and Uttar Pradesh reported separate rates for rural and urban areas, but the two states provided data only for rural areas. The rate for Punjab was reported with and without government subsidy, but the figure includes only prices with government subsidy. In Karnataka, farmers are not charged at all for electricity. The cost of supply is the weighted-average cost of supply across all sectors. kWh = kilowatt-hour.

by 0.3 percent, and the gap between expenditure and revenue without subsidies rises by 0.4 percent, all else being equal.

Besides increasing fiscal burdens, electricity subsidies to farmers have the unintended consequence of triggering an overexploitation of groundwater. Empirical evidence shows that farmers are price-sensitive in their use of irrigation water. When the cost of water extraction is artificially low, farmers are less likely to adopt water-conserving irrigation technologies and more likely to shift to water-intensive crops such as rice. Using panel data for 1995–2004 from 370 districts across India, Badiani and Jessoe (2013) find that a 10 percent increase in the average subsidy would lead to a 6.6 percent increase in extraction. Nationally, the area irrigated by groundwater has increased by more than 700 percent since fiscal 1951.

What are the welfare effects of groundwater overexploitation induced by electricity subsidies? To quantify these effects, the ideal approach is to estimate the optimal path of groundwater extraction and then estimate the welfare loss from deviating from that path. However, such an exercise would be extremely data-intensive. It would require information on the stock of groundwater, the costs of extraction, the rate of groundwater replenishment, the marginal productivity of groundwater in crop production, and so on. Rather than making assumptions about unknown variables, the analysis instead estimates how much water would be saved if the electricity price for agriculture were cost-reflective, and then it applies the shadow price of water to quantify the monetary value of groundwater overexploitation.

According to Badiani and Jessoe (2013), when the price of electricity charged to farmers increases by 1 percent, the volume of groundwater pumped falls by 0.13 percent. Using this price elasticity along with state-level data on the implicit electricity price for farmers and the pumped volume of groundwater, the analysis estimates that raising the implicit price to a cost-reflective price would reduce groundwater extraction by 86.2 billion cubic meters a year (the analysis excludes Karnataka, where farmers are not charged for electricity). The largest reductions would be expected in Haryana, Punjab, Uttar Pradesh, and Tamil Nadu. Hussain and others (2009) estimate that the shadow price of water in agriculture in Pakistan is $0.01–$0.05 per cubic meter. Using the middle value of $0.03 as an approximation for the shadow price in India, the analysis estimates that the economic cost of groundwater overexploitation is roughly $2.59 billion (0.12 percent of GDP) a year.

It should be noted that central and state governments have taken a number of corrective steps to reverse the trend of groundwater depletion. For example, Gujarat implemented feeder segregation to restrict agricultural power supply to 8 hours per day. West Bengal introduced time-of-day meters, replaced flat tariff with pro-rata tariff, and increased the price of water. These reforms have helped reduce power subsidy to agriculture and contain groundwater draft (Mukherji, Shah and Verma 2010).

Also notable is that subsidized electricity is not the only contributor to groundwater exploitation. Millions of diesel operated tube wells and solar irrigation pumps have added to groundwater strain. Lack of alternative water sources also forced farmers to rely on groundwater to meet water needs in different cropping seasons. To prevent ground water over-exploitation, other measures, such as raising awareness among farmers about crop-water budgeting, fixing minimum support price for crops, and assure availability of alternative source of water, are also needed as part of a holistic policy response.

Downstream

India has made remarkable strides in extending the electricity grid to the country's vast population. Much progress has been made through its national electrification program, Rajiv Gandhi Grameen Vidyutikaran Yojana (RGGVY), launched in 2005, and the Saubhagya Scheme, launched in 2017. The RGGVY program consolidated all rural electrification efforts into a broad commitment to electrify all 100,000 unelectrified villages of more than 100 persons and provide free electricity connections to more than 21 million rural households living below the poverty line. The Saubhagya scheme aims to complete the electrification process by December 2018. The latest statistics show that these programs have connected 286 million people to the grid and that India's household electrification rate rose from 67 percent in 2005 to 86 percent as of October 2017 (Government of India 2018).

Yet another 178 million people in India remain without access to the grid. Most are in rural areas. The access rate has reached 98 percent for the urban population but only 81 percent for the rural population. Access is also uneven across different parts of the country. According to the latest government rural electrification data, six states and the rural vicinity of the national capital, New Delhi, had achieved a 100 percent electrification rate as of October 2017 (Figure 4.24). But several states in east, north, and northeast India, including Jharkhand, Assam, Uttar Pradesh, Nagaland, and Odisha, have electrification rates below 63 percent.

For those who do have access to electricity, outages are frequent, especially in rural areas. According to the latest national household survey (Indian Human Development Survey 2011/2012) that provides information on the reliability of electricity supply, almost 45 percent of rural households connected to the grid reported power outages of at least 13 hours a day in 2011–12. The frequency and duration of power cuts may have been largely reduced over the last few years because of the significant reduction in power shortages, but no recent data on the reliability of electricity supply are available.

Access to reliable electricity is a prerequisite for sustainable social and economic development. Low access and poor quality of electricity supply would have a significant effect on households' living standards and firms' operation and growth.

FIGURE 4.24 Household electrification rates are low in east and northeast India

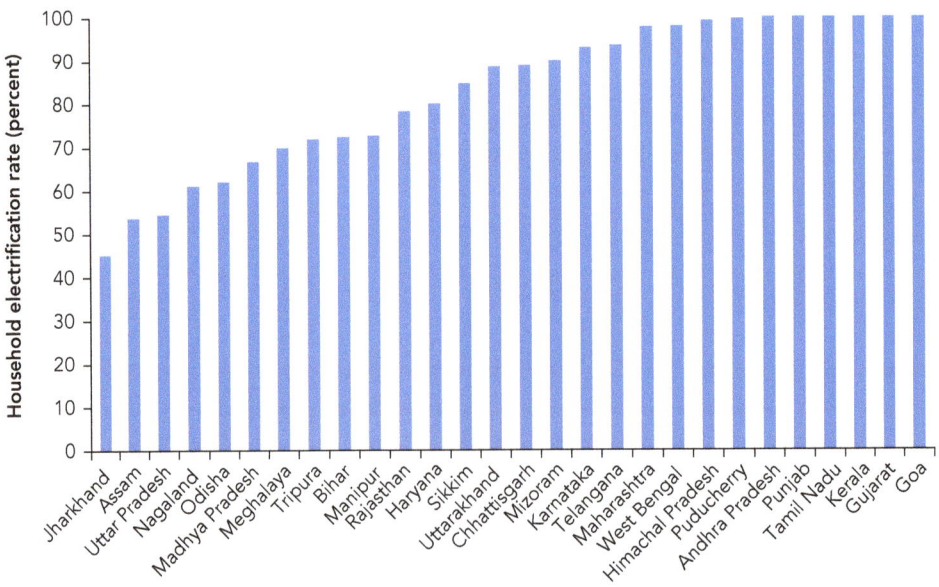

Source: Saubhagya Dashboard, http://saubhagya.gov.in/.
Note: Data reflect electrification rate as of October 11, 2017.

INSTITUTIONAL: WELFARE LOSS FOR HOUSEHOLDS

To quantify the cost to households of lack of access to electricity, the analysis uses data from the Indian Human Development Survey, which interviewed a nationally representative sample of 40,000 households during two rounds. The first round, conducted in 2004–05 (mostly in 2005), collected information on 41,554 households in 33 states and union territories, 383 districts, 1,503 villages, and 971 urban blocks. The second, conducted in 2011–12 (mostly in 2012), reinterviewed 83 percent of the original households and split households (if located within the same village or town) and interviewed 2,134 new households, for a total of 42,152 households, including more than 20,000 rural households.

Several earlier studies examined the welfare effects of grid connections on households in India (Banerjee and others 2015; Burlig and Preonas 2016; Chakravorty, Pelli, and Marchand 2014; Khandker and others 2014; van de Walle and others 2015). With the exception of Chakravorty, Pelli, and Marchand (2014), these analyses focused on estimating the causal relationship between household welfare and a binary variable of electricity access without taking into full account whether the "connected" household actually received an adequate level of service. This analysis takes advantage of the Indian Human Development Survey, the latest survey that provides nationally representative information on quality of electricity supply at the household level, to gauge the causal relationship between access to reliable electricity and household welfare. This relationship is then applied to power shortage data in fiscal 2016 to arrive at an estimation of the cost of lack of reliable access to electricity in rural India. The analysis differs from that by Chakravorty, Pelli, and Marchand (2014) in that it uses a panel rather than a cross-sectional survey and examines the effect of electrification on a broad range of households' social and economic outcomes, including expenditure, education, employment, and poverty status.

The Indian Human Development Survey asked detailed questions on household energy consumption behavior such as fuel use, cash expenditure for fuels, time spent collecting biomass fuels, and types of stoves and electrical appliances used in the household. It also asked about the average daily duration of power outages as a measure of reliability of power supply and the source of household electricity, as well as about key features of the villages in which the surveyed households were located.

Household survey data suggest that economic outcomes are better for households with access to electricity and for households with a better quality of electricity. This observational evidence comes from two simple averages: (1) a comparison of households with and without electricity and (2) a comparison of grid-connected households that have access to electricity for at least 20 hours a day with grid-connected households that have access for less than 20 hours a day. The results show that, compared with their counterparts, grid-connected households and households with more hours of electricity supply consume less kerosene (Figures 4.25 and 4.26). They also have higher incomes and expenditures and a lower poverty rate, and both boys and girls in these households spend more time studying. Meanwhile, employment hours for women increase as they gain better access to electricity. The trend in outcomes also differs between grid and

FIGURE 4.25 Households with access to electricity have better welfare outcomes than households without access in India

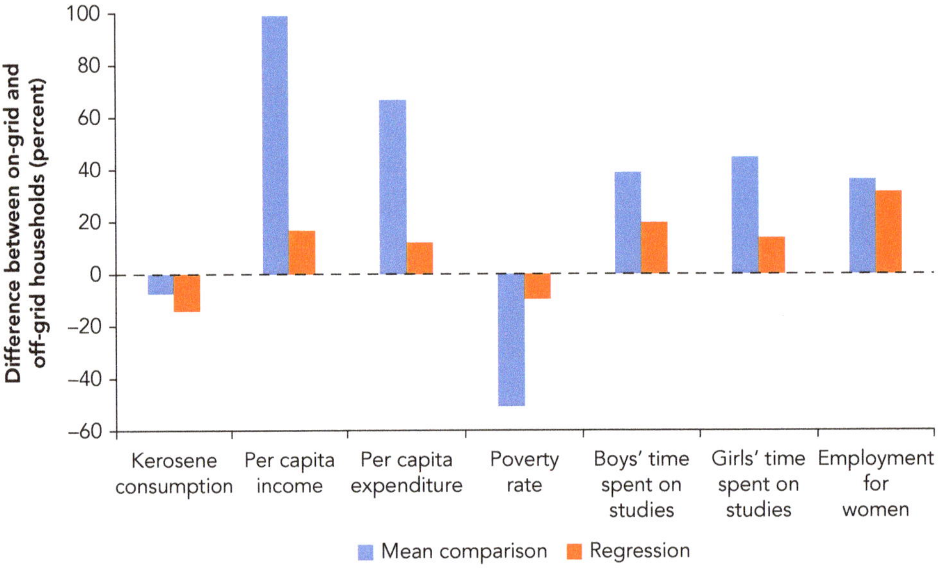

Source: Estimation based on Indian Human Development Survey (2005, 2012).
Note: Mean comparison refers to a simple average comparison without controlling for confounding factors. Regression refers to the estimated difference based on econometric analysis.

off-grid households. For example, kerosene consumption drops over time for all households, but it drops more for grid-connected households.

To what extent are the changes in household welfare caused by electrification? The analysis explores the causal relationship using econometric methods. This exercise is not straightforward because grid expansion and households' decisions to adopt electricity are typically not random and are subject to preexisting differences. For example, a government may target areas that are more easily accessible and have greater growth potential for electrification projects. And when electricity becomes available in a village, more well-off households are more likely to obtain a grid connection first.

To control for unobserved preexisting differences between connected and unconnected households, the analysis uses a two-stage propensity score–weighted fixed-effects model (Arulampalam, Booth, and Taylor 2000; Chamberlain 1984; Heckman 1981). It first estimates a household's probability of being connected to the grid (its propensity score) on the basis of a range of household and village characteristics observed in the base year (2005). It then assigns each household a weight proportional to its propensity score, so that households that look more similar to connected households in the base year receive a higher weight. Finally, it estimates the effect of changes in the access to and the reliability of electricity service on changes in outcomes for the same household between 2005 and 2012. (For a more detailed discussion of the methodology and results, see Samad and Zhang 2016.)

FIGURE 4.26 Households exposed to shorter power outages had better welfare outcomes than households exposed to longer outages in India

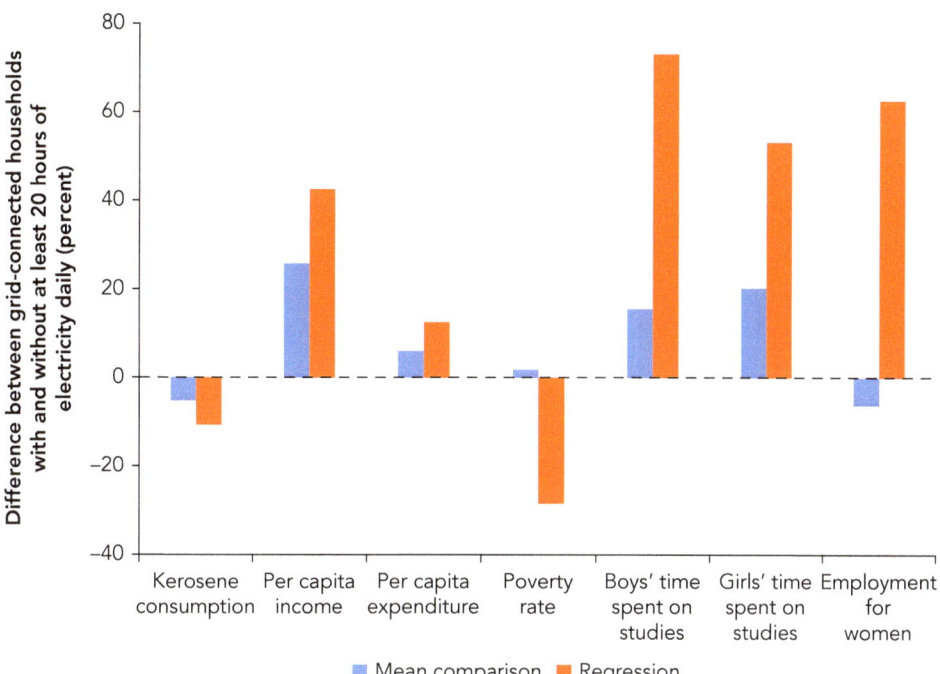

Source: Estimation based on Indian Human Development Survey (2005, 2012).
Note: Mean comparison refers to a simple average comparison without controlling for confounding factors. Regression refers to the estimated difference based on econometric analysis.

In addition, the analysis controls for a range of confounding factors at the household and village levels that may explain differences in household outcomes. They include the age, gender, and education of the head of household; the number of adult men and women in the household; the amount of household agricultural land; and measures of the household's sanitation status such as access to running water, a flush toilet, and a separate kitchen. At the community level, factors include whether a village has paved roads, a school, a market, a bank, nongovernmental organizations, and development programs and what the village prices are for alternative fuels (firewood, kerosene, liquefied petroleum gas) and essential food items (staples, meat, fish, vegetables).

The results reveal that electrification is associated with a broad range of social and economic benefits but that the extent of the benefits depends critically on the reliability of electricity service (Figures 4.25 and 4.26). Each one-hour increase in daily power outages is associated with a 2 percent reduction in households' per capita nonfarm income and a 0.5 percent reduction in their per capita total income on average. Gaining access to the grid is associated with a 16.7 percent increase in income between 2005 and 2012 when the power supply was reliable and only a 9.6 percent increase when the power supply was not.

Effects on the household poverty rate also depend on the reliability of the power supply. Gaining access to the grid is associated with a reduction in the household poverty rate of 9.5 percent when the power supply is reliable and just 6.8 percent when it is not. An additional hour of daily outages is associated with an increase in the poverty rate of 0.2 percent.

Among the long-term benefits of electricity is its potential to improve households' education outcomes. Given grid access and no power outages, boys' study time increases by one hour and girls' study time by 0.64 hours a week. Unsurprisingly, the effect of outages on all these outcomes is both negative and statistically significant. Every additional hour of daily outages is associated with a reduction in weekly study time of 0.05 hours for boys and 0.02 hours for girls. As a result, the effects of electrification are much smaller when the power supply is unreliable: boys' study time increases by 0.27 hours (about 16 minutes) and girls' study time by 0.30 hours (about 18 minutes) a week. Neither increase results in a significant improvement in grade completion.

For women, electrification helps free up time traditionally spent on collecting fuel, allowing them to spend more time on other productive activities. Gaining access to the grid has a significant positive effect on women's hours of employment, increasing them by 31 percent when the power supply is reliable and by 12 percent when it is not. Every additional hour of daily outages reduces women's hours of employment by 1.4 percent on average. Access to electricity also increases women's empowerment, measured by their decision-making ability, mobility, financial autonomy, and social participation (Samad and Zhang 2018).

To quantify the cost of lack of access to electricity, the analysis focuses solely on income loss. Assume all of the unelectrified population is in rural areas and their average per capita annual income was $130 in fiscal 2016. On the basis of an estimated income gain of 16.9 percent a year, the per capita income increase associated with obtaining access to electricity is about $22 a year. Connecting to the grid the roughly 178 million people who remained off-grid would have raised income by $3.9 billion a year.

Improving the reliability of the electricity supply would add to these gains. According to the Indian Human Development Survey, the average daily duration of outages for households with less than 24 hours of power supply was about 12.5 hours in 2012. The officially estimated power shortage declined by 75.3 percent between fiscal 2012 and fiscal 2016, from 8.5 percent to 2.1 percent. The analysis assumes that the daily duration of outages declined by the same percentage, to 3.1 hours in fiscal 2016. With an estimated income loss of −0.5 percent associated with every hour of daily outage, increasing the supply of electricity to 24 hours a day would lead to an estimated annual income gain for the rural population of $5.5 billion.

The Indian government plans to achieve 100 percent electrification in rural areas by 2019. Its plans call for connecting the villages that remain without electricity and improving the reliability of power supply in villages that are already connected. The total investment cost, covering the poles, lines, cables, transformers, and related costs, is estimated at about $4.4 billion. If this equipment is assumed to last about 30 years, the average annual investment cost would be about $147 million. In addition to initial investment costs, the marginal cost associated with electricity generation and transmission is assumed to be about $0.07 per kWh. So, even without accounting for the benefits associated with better health and education outcomes, the analysis suggests that providing reliable electricity to rural households consuming an average 900 kWh a year would produce net annual income gains of almost $7.0 billion (0.34 percent of GDP).

INSTITUTIONAL: PRODUCTIVITY LOSS FOR FIRMS

Persistent power outages impose big costs on firms. They force firms to rely on diesel generators, which produce power at a cost many times the price of electricity from the grid. Power outages, especially unscheduled ones, have an even greater impact on firms without generators because they must close down and send workers home. Unsurprisingly, nearly half of businesses in India identify lack of access to reliable power as a major constraint to their operation and growth, according to the 2014 World Bank Enterprise Survey (World Bank 2014).

Using data from the Annual Survey of Industries of India, a recent study quantifies the impact of blackouts on the revenue of manufacturing firms (Allcott, Collard-Wexler, and O'Connell 2016). It finds that power shortages of about 7.3 percent reduce revenue by 5.6–7.7 percent a year on average for firms in the survey sample. Business managers report similar effects when responding to a question in the World Bank Enterprise Survey on their firm's annual losses in sales volume as a result of shortages in electricity. In 2014, when the average electricity shortage was estimated at about 4.2 percent, managers in the manufacturing sector reported average losses of 4.9 percent of sales, and managers in the services sector reported average losses of 2.7 percent of sales.

Because both of these surveys target registered firms with at least five workers, the results do not reflect the cost of blackouts to micro- and small enterprises (MSEs), many of which are informal. Large firms can invest in generators or outsource the production of electricity-intensive intermediate goods. But MSEs have fewer options for coping with blackouts—and therefore often have to bear the full brunt of power outages. Moreover, more than 60 percent of MSEs are located in rural areas, where outages are far more frequent. The impact of power outages on MSEs is thus likely to be much larger than the impact reported in the literature.

Despite their small size, MSEs are an engine of economic growth in India. According to Hsieh and Klenow (2014), establishments in the informal sector account for about 75 percent of total manufacturing employment in India. A World Bank study (2015) estimates that MSEs may account for as much as 45 percent of India's GDP.

To better understand the effects of electricity shortages on MSEs, this analysis examines data from the Fourth All-India Census of Micro, Small and Medium Enterprises for fiscal 2005–07. Its sample includes about 1.2 million services and manufacturing enterprises, both registered and unregistered, with an average employment size of two workers. By comparison, the average employment size is 167 workers in the Annual Survey of Industries sample and 52 in the World Bank Enterprise Survey sample.

The effects of power outages on firms cannot be inferred by directly comparing the costs and profits of firms facing different levels of shortages. The severity of shortages and their impacts run in both directions. For example, more productive firms are likely to be attracted to locations with better infrastructure. Or, as Allcott, Collard-Wexler, and O'Connell (2016) note, regions with faster economic growth have a higher demand for electricity, which could in turn result in worse outages. In both cases, the effects of power outages would be underestimated. In addition, state policies may affect both infrastructure spending and the business environment.

To address simultaneity concerns, the analysis follows Allcott, Collard-Wexler, and O'Connell (2016) by using precipitation-induced variation in hydro-based power generation over time within a state as a source of exogenous variation in power shortages. As the authors show, this instrumental variable is not correlated with shocks to electricity demand at the state level, but it significantly affects shocks to electricity supply and is therefore directly correlated with power shortages at the state level.

The results reveal that the effects of electricity shortages on MSEs are almost two orders of magnitude greater than those on the larger firms reported by Allcott, Collard-Wexler, and O'Connell (2016). A 1 percentage point increase in shortages raises the production costs of larger firms by less than 0.018 percent of revenue, whereas it raises the production costs of MSEs by 0.29 percent of revenue. The effects also vary across sectors, largely reflecting differences in energy intensity. The biggest effects are in machinery manufacturing, followed by textiles. Across the full population of MSEs in the Fourth All-India survey sample, the total increase in production costs is estimated at $3.8 billion a year in fiscal 2007. (For more details on the methodology and results, see Grainger and Zhang 2017.)

Moving out again to a broader focus, the analysis looks at the welfare impact of power shortages on business as a whole. To estimate this impact, the analysis considers changes in firms' value added. The manufacturing sector in India accounted for $320 billion in value added in fiscal 2016 and the services sector for $1,107 billion. Because value added of MSEs are not separately reported, the analysis applies the estimate of the impact of power shortages on firms by World Bank Enterprise Survey for firms of all size of categories. Although the World Bank estimates are more conservative compared to Allcott, Collard-Wexler, and O'Connell (2016), it could nonetheless underestimate the impact of shortages on MSEs. Subject to that caveat, based on the officially estimated peak demand shortage of 2.1 percent of demand in fiscal 2016, total annual losses in sales as a result of electricity shortages would amount to $7.8 billion in manufacturing and $14.9 billion in services. Although manufacturing is more electricity-intensive and therefore more vulnerable to power shortages, the impact is greater in the services sector because of its greater importance in the economy. The combined losses of manufacturing and services firms amount to 1.09 percent of GDP a year.

For a couple of reasons, this calculation almost certainly underestimates the welfare impact of power shortages on firms. First, although revenue loss captures the effects of power shortages on both the quantity of production and the quality of output (through the impact on the price of products), it does not reflect the effects on job creation.

Second, since firms are likely to outsource the production of electricity-intensive intermediate inputs when electricity supply is unreliable, revenue losses from lack of reliable access to electricity are likely to be smaller than losses in value added because outsourcing increases the cost of intermediate goods (Fisher-Vanden, Mansur, and Wang 2015).

REGULATORY: CROSS-SUBSIDIES PENALIZING COMPETITIVENESS

Firms in India confront not only a low-quality electricity supply but also a high-cost one because the government cross-subsidizes residential and agricultural consumers by overcharging industrial and commercial users. Even though large industrial users are less costly to serve, the average industrial tariff was nearly twice the average residential tariff and roughly 12 percent higher than the average cost of power supply during fiscal 1993–2016 (Figure 4.27). In fiscal 2016, it was 4.2 times the average tariff for agricultural users.

Cross-subsidies impose an implicit tax on electricity for industries. And, because electricity is used to produce almost all goods and services, a tax on electricity will likely affect prices and outputs for all industries as well as foreign trade. The net effect of this tax depends on how easily consumers can substitute between different products; how easily firms can adjust the intensity of use of different production factors (such as labor and capital, which are generally complementary inputs to electricity); and how big the preexisting tax distortions are in the capital and labor markets.

FIGURE 4.27 Industrial and commercial users face higher-than-cost tariffs

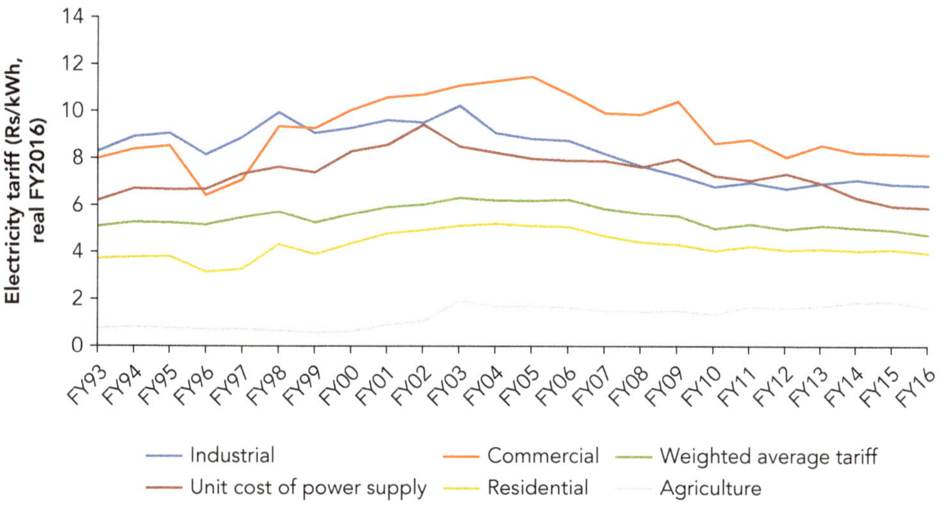

Source: Power Finance Corporation (2003, 2007, 2017) and Indian Planning Commission Power and Energy Division (2000, 2001, 2002, 2012, 2014, 2015).
Note: Unit cost of power supply for fiscal 2015 and 2016 are based on annual plan projection. Domestic prices for fiscal 2002 and 2003 were not available and were interpolated. FY = fiscal year; kWh = kilowatt-hour; Rs = Indian rupees.

One way to estimate the size of the impact is to use a computable general equilibrium model. But this approach relies on many assumptions about consumer and producer behavior. This analysis relies instead on econometrically estimated parameters to quantify the impact of cross-subsidization on the competitiveness of the Indian economy, recognizing that the total welfare cost of cross-subsidization could be much larger than the partial cost that can be estimated this way.

A higher electricity price increases the cost of both direct electricity consumption (at the final stage of production) and the indirect electricity consumption embodied in intermediate inputs. Some sectors that are not energy-intensive on the basis of direct energy consumption could still be vulnerable to energy price shocks if they use energy-intensive intermediate inputs. One example is the manufacture of machinery, which relies on steel as an intermediate input. Steel production in India has five times the direct energy intensity of machinery manufacturing. An energy price shock may therefore increase the production cost of machinery by raising the price of steel.

Because of the importance of intermediate goods, the effects of a higher electricity price on export competitiveness depend on how closely a country is integrated in the global supply chain. If a country imports most of its intermediate inputs, changes in domestic energy prices will be less important for its net exports. If it produces most of its own intermediate inputs, changes in domestic energy prices will have a substantial multiplier effect on its export competitiveness, accumulated through domestic sectoral linkages. This is the case in India, where less than 20 percent of intermediate goods are imported or exported. That level of trade dependence is strikingly low compared with the average for Organisation for Economic Co-operation and Development (OECD) and some non-OECD countries (Figure 4.28).

To quantify the impact of higher electricity prices on India's manufacturing exports, the analysis first estimates how energy costs affect bilateral trade flows using multicountry input-output information for 10 manufacturing sectors in 43 countries from 1991 to 2012. Many factors affect bilateral trade flows, including fluctuations in tariff schedules, exchange rates, and regional trade agreements. In addition to these observable factors, many shocks to trade costs and demand—such as regulatory and technological changes—are unobserved and are likely to be correlated with energy costs. The barriers to trade that each country faces with all of its trading partners (the so-called multilateral resistance terms) also affect bilateral trade flows.

To account for all observed and unobservable cofounding factors, the analysis uses a two-step estimation strategy outlined by Head and Mayer (2015). This approach allows the use of fixed effects that are consistent with the theory and properly control for the multilateral resistance terms. The results show that both direct and indirect energy costs have statistically significant effects on exports. On average across sectors, a 1 percent increase in an exporter's aggregate (direct and indirect) electricity costs is associated with a 0.014–0.125 percent reduction in exports. (For details on the methodology and results, see Chan, Manderson, and Zhang 2017.)

FIGURE 4.28 Limited trade openness amplifies the effect of electricity cross-subsidization in India

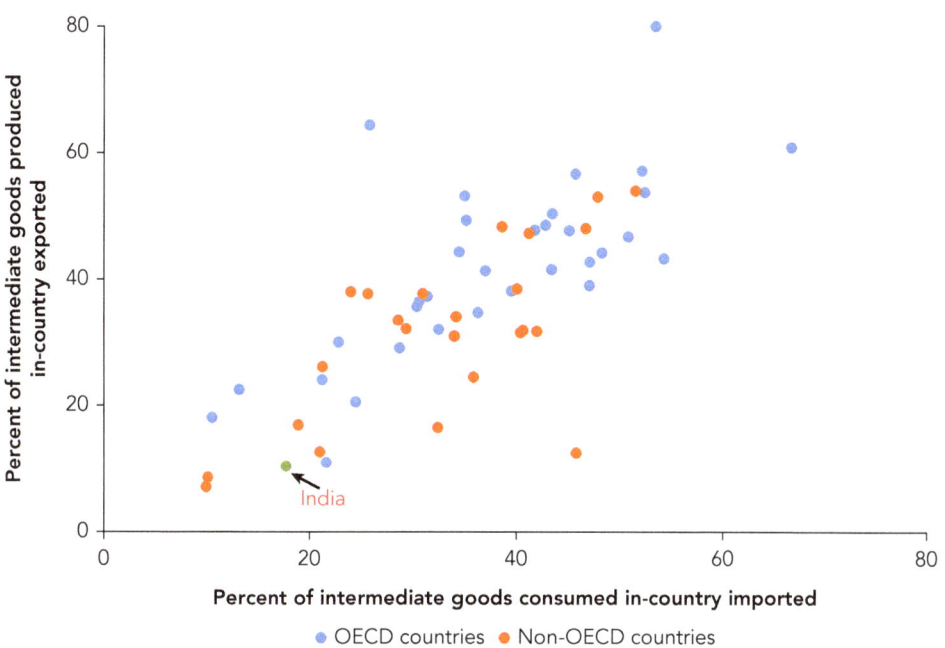

Source: Chan, Mandelson, and Zhang (2017).
Note: OECD = Organisation for Economic Co-operation and Development.

Using this estimated trade elasticity, the analysis then simulates the effect on India's net manufacturing exports of removing electricity cross-subsidies. Because India relies mostly on its own production for intermediate goods, simulations based on aggregate electricity costs show much larger effects than those based on direct electricity costs alone. The gap between the current industrial tariff and the average cost of supply is 12 percent of the average cost, which can be considered as an approximation of the level of implicit tax on industrial electricity tariff. This approximation, however, most likely underestimates the implicit tax because serving large industrial and commercial consumers is generally less expensive than serving other categories of consumers. A simulation shows that removing this level of cross-subsidies could increase India's net manufacturing exports by $2.1 billion (0.1 percent of GDP) a year. The size of the effect varies across sectors, with the basic metals, machinery, chemicals, and paper and pulp industries benefitting the most (Figure 4.29).

SOCIAL: EMISSIONS AND DISEASE FROM KEROSENE LIGHTING

Lacking reliable access to electricity, households and businesses often turn to other options to meet their basic energy needs. For households, kerosene is among the most

FIGURE 4.29 Removing electricity cross-subsidies would increase net exports in India

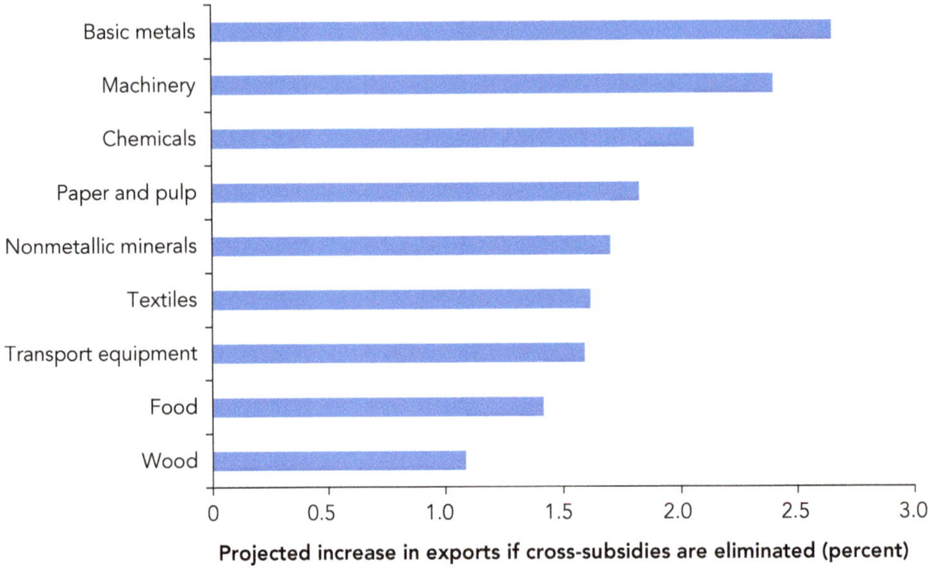

Projected increase in exports if cross-subsidies are eliminated (percent)

Source: Chan, Manderson, and Zhang (2017).

common alternatives for lighting. In India, about 165 million kerosene lamps were used by households in 2012 (Tedsen 2013).

Although largely affordable and accessible, kerosene lighting is a significant source of black carbon, a potent global warming agent. Because of its heavy reliance on kerosene lamps, India has among the highest warming (radiative forcing) effects from emissions of black carbon in the world (Lam and others 2016).

Lam and others (2016) estimate that roughly 4.9 percent of the kerosene burned in India is converted into black carbon emissions. According to the Indian Human Development Survey, in 2005 households used on average 4,670 gigagrams (Gg) of kerosene for residential lighting, 64 percent as the primary source of lighting and 36 percent for supplemental lighting. Considering that the access rate to electricity increased from 67 percent in 2005 to 85 percent in 2016, total kerosene consumption in 2016 is estimated to be 3,681 Gg. This usage implies total black carbon emissions of 181 Gg in 2016. Using a conversion rate based on the climate warming effects of black carbon and carbon dioxide (Jacobson and others 2013), estimates suggest that this amount is equivalent to about 161 million tons of carbon dioxide. Given a shadow price of $40 per ton for carbon dioxide emissions, the environmental cost of kerosene-based domestic lighting in India can be estimated at roughly $6.4 billion a year.

Whereas households cope with electricity shortages by resorting to kerosene lamps, businesses develop captive generation. According to the World Bank Enterprise Survey, in 2014 almost half of industrial and commercial establishments in India owned or

FIGURE 4.30 Captive power plants—which are generally smaller and run less frequently than utility plants—are expanding rapidly in India

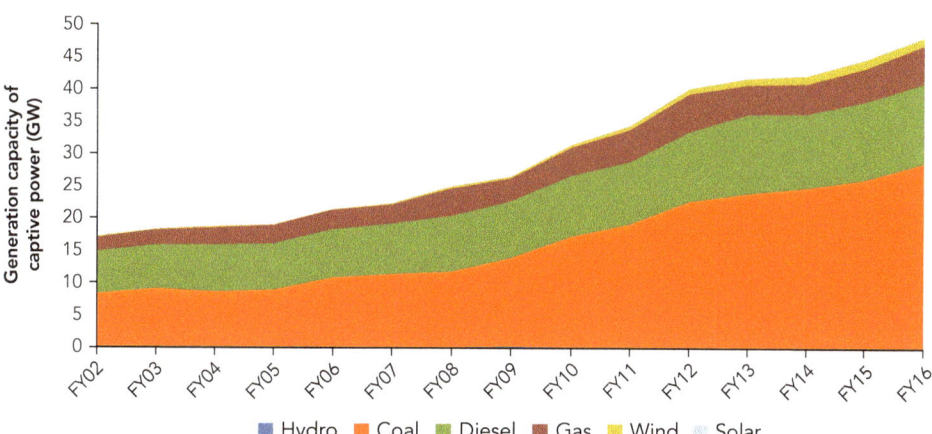

Source: Ministry of Power, India, accessed through Indiastat.
Note: FY = fiscal year; GW = gigawatt.

shared a captive power generator. The installed capacity of captive power plants has increased rapidly in recent years (Figure 4.30). In fiscal 2015, captive generation capacity reached 45 GW, almost 16 percent of total installed capacity (utilities) in India. But this number accounts only for captive plants in the organized sector that are larger than 1 MW and not the millions of smaller generators. Coal dominates the fuel mix for captive generation, accounting for 58 percent of the total. But diesel-based generation is also growing rapidly. The Indian Economic Survey 2015–16 reports that the total capacity of diesel generators may be growing by 5 GW a year (Ministry of Finance, India 2016).

Captive generation is likely to have greater environmental impacts than utility power plants relative to the amount of electricity produced. Captive plants are typically used for backup generation, tend to be smaller, and have a lower load factor. For coal plants, smaller capacity and a lower load factor are associated with a higher operating heat rate (Figure 4.31). Captive plants are therefore likely to have a greater emission intensity than larger and more frequently run utility plants. Probably more important, small captive plants are less likely to install emission control equipment and are widely dispersed across urban centers. Their emissions are therefore likely to cause much greater health damage than those of power plants that have high stacks and are located far from population centers. Because of data limitations, an accurate estimate of the externalities of captive power generation is not possible. An estimate of the downstream social cost, $6.4 billion a year, therefore represents a lower-end estimate of the true social cost.

FIGURE 4.31 Smaller power plants and plants that run less frequently are less efficient in India

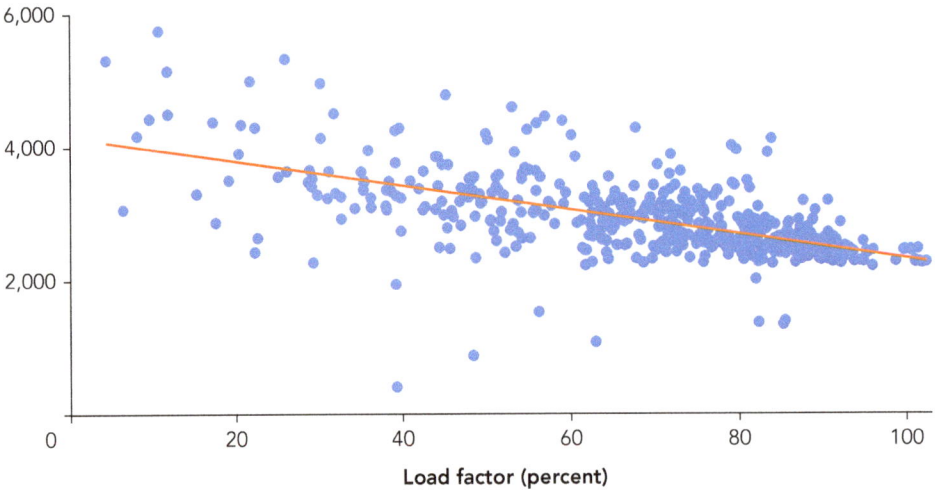

a. Operating heat rate is negatively correlated with load factor

Load factor (percent)

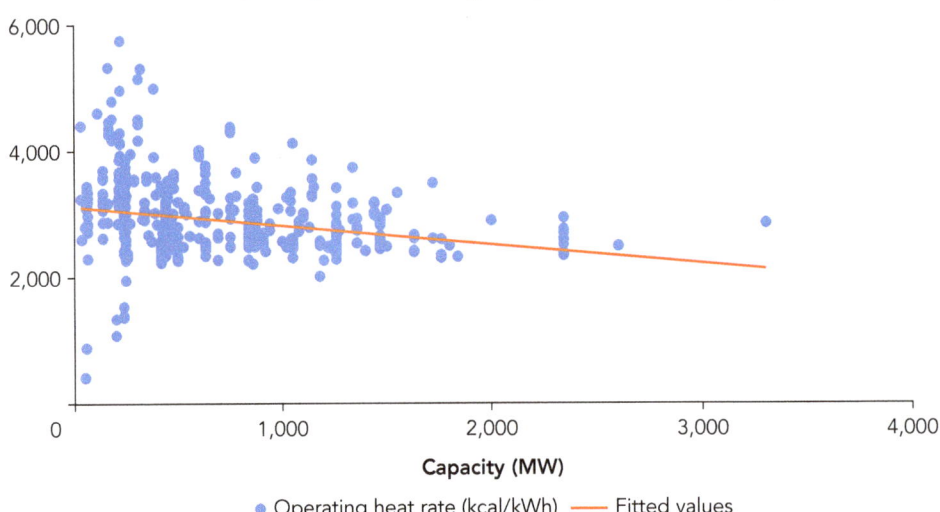

b. Operating heat rate is negatively correlated with capacity

Capacity (MW)

● Operating heat rate (kcal/kWh)   —— Fitted values

Source: Based on data from Central Electricity Authority (2000–12).
Note: kcal = kilocalorie; kWh = kilowatt-hour; MW = megawatt.

Summarizing the Costs

Distortions in the power sector imposed a total economic cost of roughly $86.1 billion (4.13 percent of GDP) on the Indian economy in fiscal 2016 (Table 4.2). The fiscal cost, consisting of subsidies to distribution utilities, was $8.8 billion (0.42 percent of GDP) in fiscal 2016 (Power Finance Corporation 2017).

TABLE 4.2 Cost of power sector distortions in India at a glance
percent of GDP

| Type of cost | Upstream | Core | | | | Downstream | Total |
		Generation	Dispatch	Transmission	Distribution		
Fiscal	0	0	0	0	0.42	0	0.42
Institutional	0.06	0.10	—	0.02	0.10	1.42	1.70
Regulatory	0.19	0	—	—	0.02	0.10	0.31
Social	1.69	0	—	—	0.12	0.31	2.12
Economic	1.94	0.10	—	0.02	0.24	1.83	4.13

Source: World Bank estimation.
Note: — = Not available. Estimation is for fiscal 2016.

The greatest source of waste is excessive coal-fired power generation, which leads to substantial health and environmental damages. The excessive health cost borne by the population is estimated at $35.9 billion a year. Excessive emissions of global warming gases add another $14.1 billion in annual external costs. Together, the net social and environmental costs of excessive coal consumption reach roughly 1.7 percent of GDP a year.

The impact of power shortages on downstream rural households and firms is the second-largest source of economic cost, estimated at 1.42 percent of GDP a year. It includes the potential income losses of unelectrified households and the income losses of households and firms that are already connected to the grid but affected by power outages. This number likely underestimates the actual cost of power shortages for two reasons. First, the estimation is based on the officially reported average demand shortages of 2.1 percent in fiscal 2016. Actual power shortages could be much larger because the estimated demand does not reflect the latent demand of the unserved or underserved population and businesses. Second, lack of reliable access to electricity has negative implications for a range of social and economic outcomes such as educational achievement and gender equality. These losses are difficult to quantify and are therefore not included in the calculation.

The third-largest cost is downstream social distortions from the use of kerosene lamps, which are estimated to cost the economy 0.31 percent of GDP. These distortions are followed by (1) regulatory distortions upstream (including the underpricing of coal and the cross-subsidization of passenger railway service from freight, both of which exacerbate coal shortages, resulting in a combined welfare loss of 0.19 percent of GDP); (2) groundwater depletion induced by electricity subsidies (0.12 percent of GDP); (3) inefficient electricity generation and distribution, both of which cost the economy an estimated 0.1 percent of GDP a year; and (4) electricity cross-subsidies, which undermine the international competitiveness of manufacturing (0.1 percent of GDP).

This chapter provides only a qualitative discussion of the social cost of coal mining and captive power generation because the data needed for an economic valuation of these impacts do not exist. In addition, nonperforming power sector loans threaten the stability of India's financial sector (Pargal and Banerjee 2014). Their impact on economic growth was not quantified due to lack of data, however. Some of the analysis, including plant-level analysis on generation efficiency, is based on data from fiscal 2000–12, the latest year in which data were available at the time of the analysis. Simulation on potential increase in electricity supply from improving generation efficiency is based on data for fiscal 2015. Overall, for various reasons described in the chapter, the estimate of the total economic cost of distortions may represent a lower-bound estimate of the actual cost.

Notes

1. For example, Reliance Sasan Ultra Mega Power Plant which is located near captive coal blocks Mohan and Mohar-Amlohri Ext has a levelized tariff of Rs. 1.196 per kWh as compared to the average electricity tariff of Rs 3.5–4 per kWh.

2. Under flat-rate tariffs, consumers pay a fixed monthly fee based on the horsepower rating of the electric motor used rather than paying per kWh. On the basis of the pump size used, the regulator can determine a monthly fixed charge to achieve a flat implicit price per kilowatt-hour. Pump size is an indication of electricity drawn per month. The implicit rate is a function of the fixed charge and the estimated number of kilowatt-hours that will be drawn by the pump. It is possible for the implicit tariff to be the same across pump sizes, but that does not mean that the fixed charge is also the same across pumps. If, for example, the fixed charge for a pump that uses 200 kWh per month is half the flat rate for a pump that uses 400 kWh per month, the implicit rate per kWh is the same (Badiani and Jessoe 2013).

References

Allcott, H., A. Collard-Wexler, and S. D. O'Connell. 2016. "How Do Electricity Shortages Affect Industry? Evidence from India." *American Economic Review* 106 (3): 587–624.

Arulampalam, W., A. L. Booth, and M. P. Taylor. 2000. "Unemployment Persistence." *Oxford Economic Papers* 52: 24–50.

Australian Department of Industry and Science. 2015. *Coal in India*. Canberra.

Badiani, Reena, and Katrina Jessoe. 2013. "The Impact of Electricity Subsidies on Groundwater Extraction and Agricultural Production." Working paper, University of California, Davis.

Banerjee, Sudeshna Ghosh, Douglas Barnes, Bipul Singh, Kristy Mayer, and Hussain Samad. 2015. *Power for All: Electricity Access Challenge in India*. Washington, DC: World Bank.

Burke, Paul J., and Hua Liao. 2015. "Is the Price Elasticity of Demand for Coal in China Increasing?" *China Economic Review* 5 (36): 309–22.

Burlig, Fiona, and Louis Preonas. 2016. "Out of the Darkness and into the Light? Development Effects of Rural Electrification in India." Energy Institute at Haas Working Paper 268, Haas School of Business, University of California at Berkeley.

Central Electricity Authority. 2000–2012. Performance Review of Thermal Power Stations. New Delhi.

———. 2017a. *Executive Summary for the Month of April 2017*. Ministry of Power, Government of India, New Delhi. http://www.cea.nic.in/reports/monthly/executivesummary/2017/exe _summary-04.pdf.

———. 2017b. *Growth of Electricity Sector in India from 1947–2017*. Government of India, New Delhi.

———. 2018. *Load Generation Balance Report 2018–2019*. Government of India, New Delhi.

CERC (Central Electricity Regulatory Commission). 2015. *Annual Report 2014–2015*. New Delhi.

———. 2017. *Report on Short-Term Power Market in India: 2016–17*. Economics Division, New Delhi.

Chakravorty, Ujjayant, Martino Pelli, and Beaza P. Ural Marchand. 2014. "Does the Quality of Electricity Matter? Evidence from Rural India." *Journal of Economic Behavior and Organization* 107: 228–47.

Chamberlain, G. 1984. "Panel Data." In *Handbook of Econometrics*, edited by S. Griliches and M. Intriligator. Amsterdam: North-Holland.

Chan, Ron, Edward Manderson, and Fan Zhang. 2017. "Energy Prices and International Trade: Incorporating Input-Output Linkages." Policy Research Working Paper 8076, World Bank, Washington, DC.

CIL (Coal India Ltd.). 2016. *Annual Report 2015–2016*. New Delhi.

———. 2018. "Price Notification." New Delhi. https://www.coalindia.in/DesktopModules /DocumentList/documents/Price_Notification_dated_08.01.2018_effective_from_0000_Hrs _of_09.01.2018_09012018.pdf.

Confederation of Indian Industries (CII). 2011. "Stakeholder Analysis Report on Perform, Achieve and Trade (PAT) Scheme of Government of India."

Dahl, Carol A. 1993. *A Survey of Energy Demand Elasticities in Support of the Development of the NEMS*. Golden: Colorado School of Mines.

Directorate General of Mines Safety. 2005. *Annual Report*. Ministry of Labor and Employment, Government of India, New Delhi.

———. 2009. *Annual Report*. Ministry of Labor and Employment, Government of India, New Delhi.

———. 2014. *Statistics of Mines in India*. Vol. 1 (Coal). Ministry of Labor and Employment, Government of India, New Delhi.

Downing, T. 2002. "Avoiding New Poverty: Mining-Induced Displacement and Resettlement." World Business Council for Sustainable Development, Geneva.

EIA (Energy Information Administration). 2015. "India's Coal Industry in Flux as Government Sets Ambitious Coal Production Targets." Washington, DC.

———. 2016. *Annual Coal Report 2015*. Washington, DC.

Epstein, Paul. R., Jonathan J. Buonocore, Kevin Eckerle, Michael Hendryx, Benjamin M. Stout III, Richard Heinberg, Richard W. Clapp, Beverly May, Nancy L. Reinhart, Melissa M. Ahern, Samir K. Doshi, and Leslie Glustrom. 2011. "Full Cost Accounting for the Life Cycle of Coal." *Annals of New York Academy of Sciences* 1219: 73–98.

Fabrizio, Kira, Nancy L. Rose, and Catherine D. Wolfram. 2007 "Do Markets Reduce Costs? Assessing the Impact of Regulatory Restructuring on US Electric Generation Efficiency." *American Economic Review* 97 (4): 1250–77.

Filippini, Massimo, and Shonali Pachauri. 2004. "Elasticities of Electricity Demand in Urban Indian Households." *Energy Policy* 32 (3): 429–36.

Fisher-Vanden, K., E. T. Mansur, and Q. J. Wang. 2015. "Electricity Shortages and Firm Productivity: Evidence from China's Industrial Firms." *Journal of Development Economics* 114: 172–88.

Global Burden of Disease MAPs Working Group. 2018. *Burden of Disease Attributable to Major Air Pollution Source in India Special Report.* Boston.

Government of India. 2018. *Saubhagya.* New Delhi. http://saubhagya.gov.in/dashboard?lang=en.

Grainger, C. A., and Fan Zhang. 2017. "The Impact of Electricity Shortages on Micro- and Small-Enterprises: Evidence from India." Background paper prepared for this report, World Bank, Washington, DC.

Grausz, Samuel. 2011. *The Social Cost of Coal: Implications for the World Bank.* Climate Advisers, Washington, DC.

Head, Keith, and Thierry Mayer. 2015. "Gravity Equations: Workhorse, Toolkit, and Cookbook." In *Handbook of International Economics*, Vol. 4, edited by Gita Gopinath, Elhanan Helpman, and Kenneth Rogoff, 131–95. Amsterdam: Elsevier.

Health Effects Institute. 2017. *State of Global Air: A Special Report on Global Exposure to Air Pollution and Its Disease Burden.* Boston.

Heckman, James J. 1981. "The Incidental Parameters Problem and the Problem of Initial Conditions in Estimating a Discrete Time–Discrete Data Stochastic Process." In *Structural Analysis of Discrete Data with Econometric Applications*, edited by C. Manski and D. McFadden. Cambridge, MA: MIT Press.

Hsieh, C.-T., and P. J. Klenow. 2014. "The Life Cycle of Plants in India and Mexico." *Quarterly Journal of Economics* 129 (3): 1035–84.

Hussain, Ijaz, Maqbool H. Sial, Zakir Hussain, and Waqar Akram. 2009. "Economic Value of Irrigation Water: Evidence from a Punjab Canal." *Lahore Journal of Economics* 14 (1): 69–84.

IEA (International Energy Agency). 2017a. *CO$_2$ Emissions from Fuel Combustion.* Paris.

———. 2017b. World Energy Balance and Statistics Database. https://www.iea.org/statistics/relateddatabases/worldenergystatisticsandbalances/.

———. 2017c. *World Energy Outlook.* Paris.

Indiastat. Various years. Database. https://www.indiastat.com/.

Indian Planning Commission, Power and Energy Division. Various years. *On the Working of State Power Utilities and Electricity Departments.* New Delhi.

IPCC (Intergovernmental Panel on Climate Change). 2006. *Guidelines for National Greenhouse Gas Inventories.* Geneva.

Jacobson, Arne, Tami Bond, Nicholas L. Lam, and Nathan Hultman. 2013. *Black Carbon and Kerosene Lighting: An Opportunity for Rapid Action on Climate Change and Clean Energy for Development.* Policy Paper 2013-03, Brookings Institution, Washington, DC.

Joyce, S. 1998. "Major Issues in Miner Health." *Environmental Health Perspectives* 106: A538–A543.

Khandker, Shahidur R., Hussain A. Samad, Rubaba Ali, and Douglas F. Barnes. 2014. "Who Benefits Most from Rural Electrification? Evidence from India." *Energy Journal* 35 (2): 75–96.

Khurana, Mani, and Sudeshna Ghosh Banerjee. 2014. *Beyond Crisis: The Financial Performance of India's Power Sector.* Washington, DC: World Bank.

Kulshreshtha, Mudit, and Jyoti K. Parikh. 2000. "Modeling Demand for Coal in India: Vector Autoregressive Models with Cointegrated Variables." *Energy* 25 (2): 149–68.

Lam, Nicholas, L. Shonali Pachauri, Pallav Purohit, Yu Nagai, Michael N. Bates, Colin Cameron, and Kirk R. Smith. 2016. "Kerosene Subsidies for Household Lighting in India: What Are the Impacts?" *Environmental Research Letters* 11 (4): 044014.

Lawrence, Kurt, and Micah Nehring. 2015. "Market Structure Differences Impacting Australian Iron Ore and Metallurgical Coal Industries." *Mineral* 5 (3): 473–87.

McRae, S. 2015. "Infrastructure Quality and the Subsidy Trap." *American Economic Review* 105 (1): 35–66.

Min, Brian. 2016. "Why Does the Power Get Cut? Blackouts and the Democratic Quality Deficit." Proceedings, 2016 Annual Meeting, American Political Science Association.

Min, B., K. M. Gaba, O. F. Sarr, and A. Agalassou. 2013. "Detection of Rural Electrification in Africa Using DMSP-OLS Night Lights Imagery." *International Journal of Remote Sensing* 34 (22): 8118–41.

Min, Brian, and K. M. Gaba. 2014. "Tracking Electrification in Vietnam using Nighttime Lights." *Remote Sensing* 6 (10): 9511–29.

Min, Brian, Zachary O'Keeffe, and Fan Zhang. 2017. "Whose Power Gets Cut? Using High-Frequency Satellite Images to Measure Power Supply Irregularity." Policy Research Working Paper 8131, World Bank, Washington, DC.

Ministry of Coal, India. Various years. *Provisional Coal Statistics.* New Delhi.

———. 2010. *Coal Directory of India.* New Delhi.

Ministry of Commerce and Industry, India. Various years. Wholesale Price Index (database). New Delhi. http://www.eaindustry.nic.in/download_data_0405.asp.

Ministry of Finance, India. 2016. *Economic Survey 2015–16.* New Delhi.

Ministry of Labor and Employment, India. 2016. *Statistics of Mines In India, Volume-I (Coal).* New Delhi.

Ministry of Railways, India. 2017. *Indian Railways Statistical Publications 2015–16.* New Delhi.

Mukherji, A, Shah, T and Verma, S. 2010. Electricity reforms and its impact on groundwater use: evidence from India. International Water Management Institute Conference Paper.

NTPC (National Thermal Power Corporation) and Central Board of Irrigation and Power. 2016. Thermal Power Plants-Coal Linkages (map). New Delhi.

OECD (Organisation for Economic Co-operation and Development). 2013. *Recent Developments in Rail Transportation Services.* Directorate for Financial and Enterprise Affairs Competition Committee, Paris.

Pargal, Sheoli, and Sudeshna Ghosh Banerjee. 2014. *More Power to India: The Challenge of Electricity Distribution.* Washington, DC: World Bank.

Pargal, Sheoli, and Kristy Mayer. 2015. *Governance of Indian State Power Utilities: An Ongoing Journey.* Directions in Development—Energy and Mining. Washington, DC: World Bank.

Parry, Ian, Dirk Hein, Eliza Lis, and Shanjun Li. 2014. *Getting Energy Prices Right: From Principle to Practice.* Washington, DC: International Monetary Fund.

Power Finance Corporation. Various years. *Performance of State Power Utilities.* Government of India, New Delhi.

Press Information Bureau. 2017. *Year End Review 2017—MNRE.* Government of India, New Delhi.

Qazi, Parvez Akhtar, and Rita Tahilramani. 2017. *Indian Railways: Re-birth of the Colossus.* Hyderabad: Edelweiss Securities Limited.

Ryan, N. 2017. "The Competitive Effects of Transmission Infrastructure in the Indian Electricity Market." NBER Working Paper 2310, National Bureau of Economic Research, Cambridge, MA.

Safe Work Australia. 2018. "Mining Fact Sheet." https://www.safeworkaustralia.gov.au /industry_business/mining.

Samad, Hussain, and Fan Zhang. 2016. "Benefits of Electrification and the Role of Reliability: Evidence from India." Policy Research Working Paper 7889, World Bank, Washington, DC.

———. 2018. "Does Electrification Help Empowerment Women? Evidence from Rural India." Policy Research Working Paper 6095, World Bank, Washington, DC.

Schwab, Klaus. 2018. *The Global Competitiveness Report 2017–2018.* Geneva: World Economic Forum.

Sharma, Rajat, and Yogima Seth Sharma. 2016. "Rail Ministry Mulling 'Give up' Ticket Subsidy Move." *Economic Times*, June 28. https://economictimes.indiatimes.com/industry/transpor tation/railways/rail-ministry-mulling-give-up-ticket-subsidy-move/articleshow/52954378 .cms.

Singh, M. 2015. "Mining and Its Impact on Tribals in India: Socio-Economic and Environmental Risks." *International Journal of Social Science and Humanities Research* 3 (2): 429–39.

Tedsen, E. 2013. *Black Carbon Emissions from Kerosene Lamps: Potential for a New CCAC Initiative.* Berlin: Ecologic Institute.

TERI (The Energy Resources Institute). 2013. *The Energy Data Directory and Yearbook (TEDDY) 2012/13.* New Delhi.

Tiwari, Rachit, S. Bhattacharya, and Piyush Raghav. 2015. "A Discussion on Non-coking Coal Pricing Systems Adopted in Different Countries." *Journal of Decision Makers* 40 (1): 62–73.

UIC (International Union of Railways). 2012. *International Railway Statistics 2010.* Paris.

van de Walle, Dominique, Martin Ravallion, Vibhuti Mendiratta, and Gayatri Koolwal. 2015. "Long-Term Gains from Electrification in Rural India." *World Bank Economic Review* 31 (2): 385–411.

World Bank. 2013. *Multi Country Review of Key Issues in the Power Sector in South Asia.* Washington, DC.

———. 2014. Enterprise Survey. Washington, DC. http://www.enterprisesurveys.org.

———. 2015. *Small and Medium Enterprises Finance.* Washington, DC.

Zhang, Fan. 2015. "Energy Price Reform and Household Welfare: The Case of Turkey." *Energy Journal* 36 (2): 71–95.

CHAPTER 5

Pakistan

Pakistan has made great strides in building a power supply system since its independence in 1947. Generation capacity increased from 60 megawatts (MW) in 1947 to 25 gigawatts (GW) in fiscal 2017. The share of the population with access to electricity grew from less than 60 percent in 1991 to an officially estimated 98 percent in 2014. And per capita electricity consumption rose from 94 kilowatt-hours (kWh) in 1971 to 471 kWh in 2014 (World Bank, World Development Indicators, 2017). The rapid progress in the early years of power sector development improved agriculture, boosted industrialization, and enhanced living conditions.

Recently, however, Pakistan has been plagued by severe shortages of electricity. Power cuts have been widespread in both urban and rural areas. The gap between projected demand and actual supply has been steadily widening since 2006 (Figure 5.1). In 2013 the shortfall reached 26 percent of total demand. During peak hours, it averaged 5,000–6,000 MW. Utilities have responded through scheduled power outages (load shedding) that can last 6–14 hours a day. In some areas, summertime load shedding has regularly extended to 18–20 hours a day. Indeed, Pakistan's power supply is one of the most unreliable in the world, disrupting business operations and people's lives (Figure 5.2). Anecdotal evidence suggests that power cuts have been reduced over the past couple of years, thanks to additional generation capacity and low global oil prices. But the 2018 *Global Competitiveness Report* still ranks Pakistan 115th among 137 economies in the reliability of electricity supply (Schwab 2018). Per capita electricity consumption, after peaking in 2006, failed to grow for almost a decade.

In an effort to close the gap between electricity demand and supply, the government has been seeking new sources of fuel. It started importing liquefied natural gas (LNG) in 2015 and has begun mining the Thar lignite deposit. Coal has historically played a negligible role in electricity production in Pakistan, accounting for less than 1 percent

FIGURE 5.1 Pakistan faces massive electricity shortages

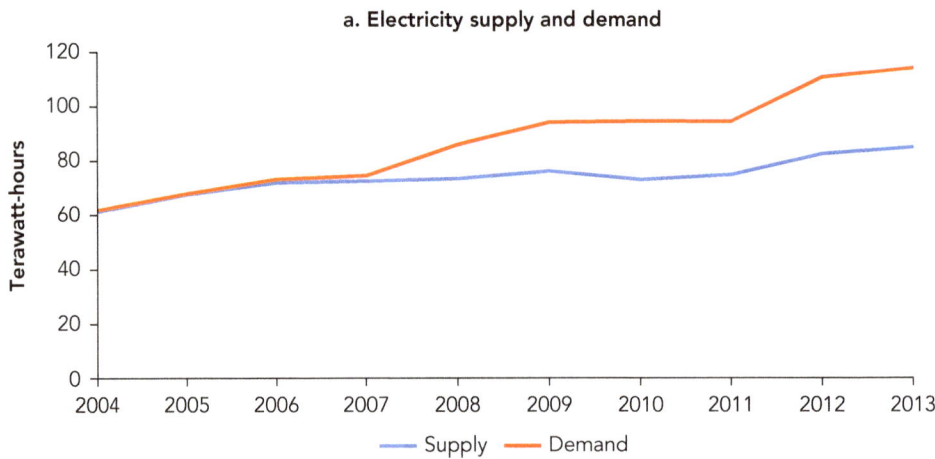

a. Electricity supply and demand

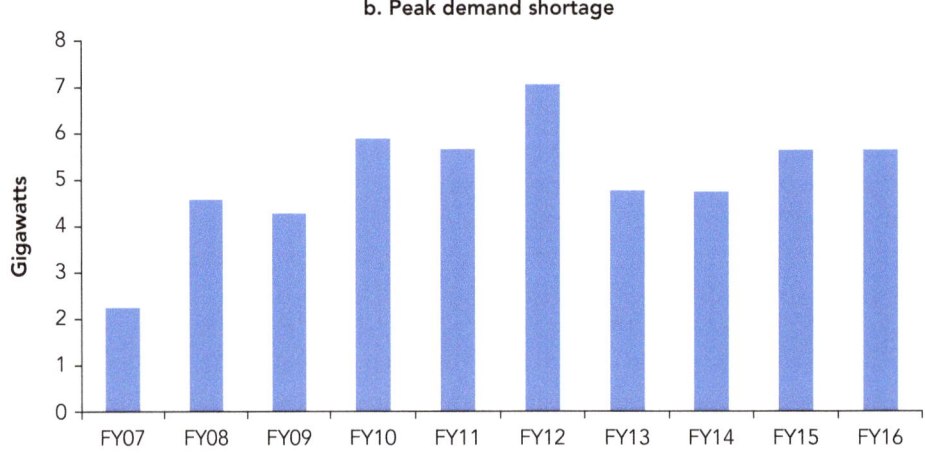

b. Peak demand shortage

Source: Panel a: electricity demand forecast: NTDC (2014); NEPRA, *State of Industry Report* (2014); panel b: NEPRA, *State of Industry Report* (2010, 2012, 2016).
Note: FY = fiscal year.

in 2014. But some government estimates suggest that potential future investments in coal-fired generation could increase that share to 30 percent.

The increased use of coal raises health and environmental concerns. Pakistan already has one of the world's highest health burdens attributable to poor air quality. One measure is the age-standardized loss of healthy life expectancy—disability-adjusted life years per 100,000 people—from exposure to fine particulate matter. The rate in Pakistan exceeds that in Bangladesh and India and is 10 times that in the United States (Health Effects Institute 2017). Greater reliance on coal-fired power generation would only worsen the situation.

FIGURE 5.2 Pakistan has one of the most unreliable power supplies in the world

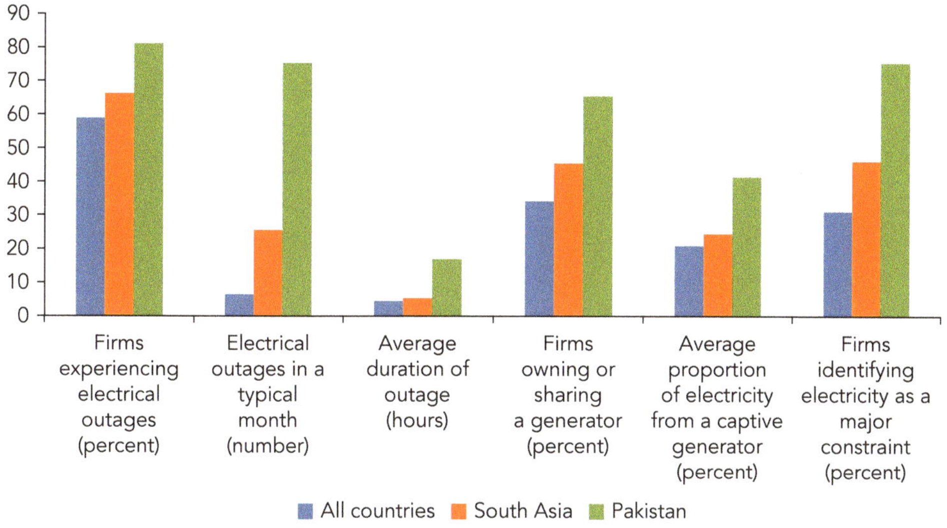

Source: World Bank (2013).

Electricity shortages in Pakistan stem from several institutional and regulatory causes. Reducing these shortages will require comprehensive sector reform aimed at addressing inefficiencies in the allocation and distribution of natural gas, increasing fuel efficiency in electricity generation, reducing losses in the transmission and distribution of electricity, and correcting pricing problems in the electricity market. The World Bank recently supported the government of Pakistan in some of these reform efforts through two Power Sector Reform Development Policy Credits. Deepening energy sector reforms would also provide opportunities to limit future reliance on coal, which would reduce both health damages and greenhouse gas emissions.

Upstream

Pakistan has sizable reserves of natural gas and coal as well as substantial hydro potential but limited known reserves of crude oil, uranium, or other nuclear fuels. Until 1973, the main thrust of the government's energy policy was therefore to expand the role of natural gas and hydropower in meeting the country's energy needs. Since 1973, the government has stepped up domestic oil exploration to reduce dependence on energy imports, while easing efforts to tap its hydro potential. The power generation policies of 1994, 2002, and 2015 maintained the focus on thermal fuel. Consequently, oil products and natural gas continue to be the largest sources of commercial energy supply in

Pakistan, together accounting for 81 percent of total energy supply and 65 percent of electricity supply in 2015 (Figure 5.3).

Natural gas is the most important indigenous fossil fuel, accounting for 50 percent of commercial energy use. In addition to electricity generation, it is used in fertilizer production and in manufacturing; for residential cooking and heating; and, as compressed natural gas, for transportation. The consumption of natural gas has increased rapidly, rising from only 9 million cubic feet per day (MMcfd) in fiscal 1955 to 4,000 MMcfd in fiscal 2015. By 2016 Pakistan was the 19th-largest user of natural gas in the world, with consumption roughly equal to that of Turkey and a quarter that of China.

The intensified use of natural gas has led to a widening gap between supply and demand (Figure 5.4). With no large new discoveries in recent years, Pakistan's gas production has stagnated at about 4,000 MMcfd. Pakistan's economic survey reports that the country's constrained demand for natural gas is roughly 6,000 MMcfd, implying a shortage of about one-third of already constrained demand (Ministry of Finance, Pakistan 2015). The government projects that, if no new supply is brought online, the gap between supply and demand could triple to 6,000 MMcfd by fiscal 2030.

The shortage of gas for electricity generation has increased reliance on expensive imported oil. With domestic gas production faltering, the government adopted a gas allocation policy in 2005 that gives lower priority to the power sector—gas goes first to residential users and the fertilizer sector. To bridge the shortfall, power generators have increasingly turned to diesel and furnace oil. Many of the oil-based plants in Pakistan were built during the 1990s, when international oil prices were at a record low. With the steep rise in global oil prices since 2002, oil has become the most expensive source of electricity generation. Some 1.2 GW of new oil-based generation capacity was nevertheless added between fiscal 2005 and fiscal 2014.

FIGURE 5.3 Natural gas and oil dominate the energy landscape in Pakistan

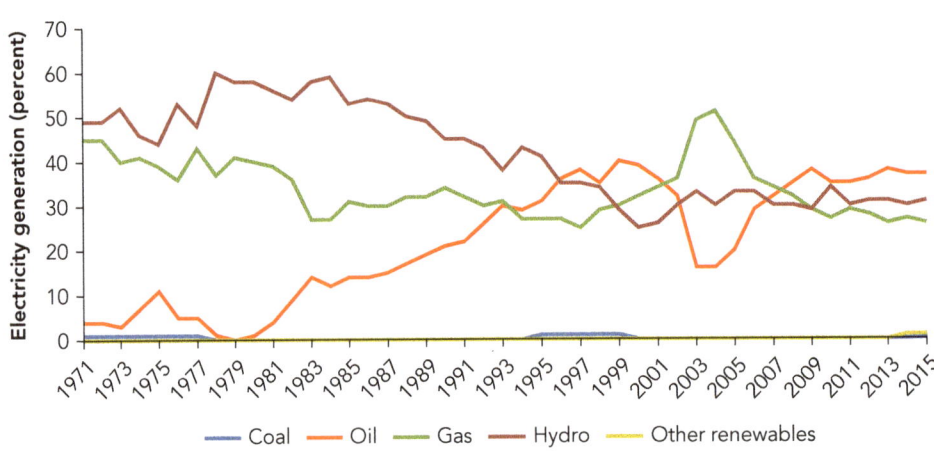

Source: IEA (2017a).

FIGURE 5.4 The shortage of natural gas has become severe in Pakistan

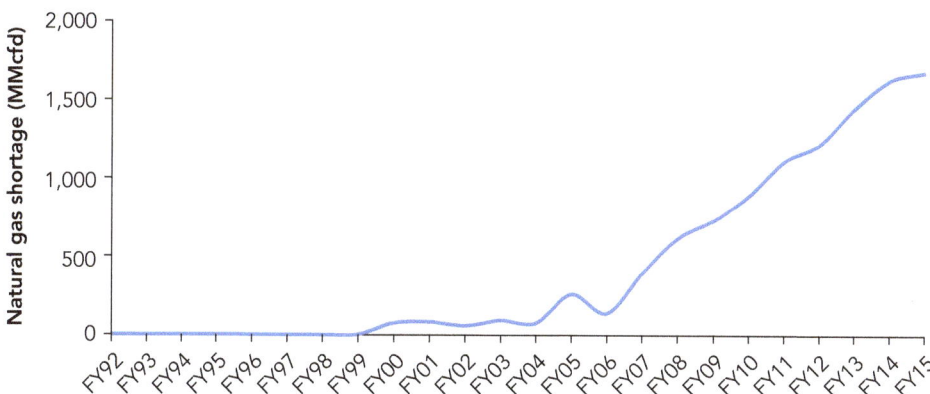

Source: Ministry of Energy, Pakistan, Division of Petroleum and Natural Resources, reproduced from Ministry of Finance (2015).
Note: MMcfd = million cubic feet per day; FY = fiscal year.

The growing reliance on oil-based generation has increased both the country's trade bills and the cost of electricity. Domestic oil production meets only 17 percent of the country's oil needs for electricity generation; the rest must be imported. In fiscal 2015, Pakistan's oil imports amounted to $11.8 billion, or about a quarter of the country's trade bill. Oil-based power generation, including furnace oil and high-speed diesel units, is much costlier than gas-based generation (Figure 5.5). Oil-based generation accounted for almost 40 percent of overall power generation in 2014 but about 70 percent of the total cost.

The high cost of generation has had a knock-on effect along the electricity supply chain. It increases the needs for subsidies to keep prices low, and when the government fails to pay subsidies on time, distribution companies cannot pay the generators. They in turn cannot pay the oil and gas suppliers, forcing generators to shut down or run at low capacity. The result is more load shedding. By some estimates, all power plants in Pakistan were running below capacity in 2014 because of fuel shortages, with oil and gas shortages alone leading to 5,000 MW of idled capacity (World Bank 2015).

Pakistan has started importing LNG since 2015, and the increased supply is expected to relieve fuel shortages and cut demand for furnace oil for power generation.

INSTITUTIONAL: GAS ALLOCATED TO UNCOMPETITIVE FERTILIZER PLANTS

In early 2000, Pakistan launched an ambitious reform program to deregulate the gas sector. Yet the government continues to maintain firm control over the supply, allocation, and pricing of gas. In 2005, facing an acute gas shortage, the government announced a policy for allocating gas among sectors of the economy. It gave first priority

FIGURE 5.5 Oil-based power plants are more expensive than gas-based power plants in Pakistan

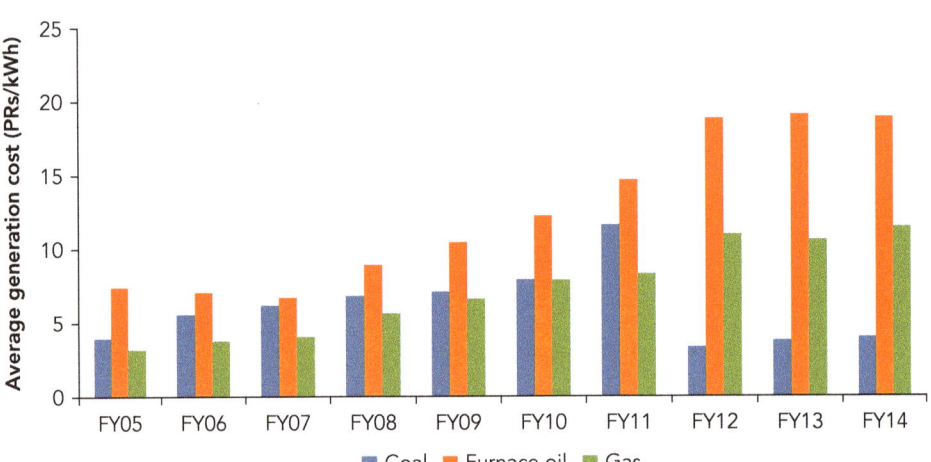

Source: Based on plant-level data, NEPRA, *State of Industry Report* (2006–15).
Note: kWh = kilowatt-hour; PRs = Pakistani rupees; FY= fiscal year.

to the residential and commercial sectors, followed by the fertilizer sector and then the power sector (Ministry of Petroleum and Natural Resources 2005). Across all sectors in Pakistan, gas could be expected to have its greatest economic benefits when used in power generation (USAID 2011). Yet, as a direct result of the policy, gas has been diverted to other sectors, and the share used for power generation fell by 33 percent during fiscal 2005–11 (Ministry of Finance, Pakistan 2013).

The government revised its gas allocation policy in 2012 with the aim of curtailing electricity supply shortages, but a large amount of subsidized gas was still reserved for lower-value uses, such as fertilizer production. The revised policy moved power generation up in the priority ranking, after the residential and commercial sectors (Ministry of Finance, Pakistan 2017). In fiscal 2016, however, about 19 percent of gas was used for fertilizer production and only 33 percent for power generation.

Misallocation of gas has an opportunity cost. To gauge this cost, the analysis considers a scenario in which the gas used by the fertilizer sector is diverted to power generation and imports replace domestically produced fertilizer—urea and DAP, two of the most commonly used types. This change increases the gas supply for power generation by 120 percent. The additional gas is allocated to power plants following a merit order, so that the most efficient ones receive gas first, up to their maximum capacity. Besides increasing the amount of electricity generated, the additional gas supply would allow dual-fuel plants to shift production to gas capacity, reducing their need for furnace oil or high-speed diesel. That said, complete dependence on fertilizer imports could lead to concerns about food supply security. The analysis does not consider the premium attached to having greater food self-sufficiency.

According to the simulation, plants' load factor (the actual energy generated as a percentage of the maximum that could be generated given their nameplate capacity) increases from 53 percent to 70 percent on average for independent power plants and from 35 percent to 45 percent for public sector generation plants with efficiency greater than 30 percent. The total amount of electricity generated rises by roughly 17 percent, from 97 terawatt-hours (TWh) to 113.5 TWh a year. Assuming that urea and DAP are imported at international prices with an additional 15 percent in local transport costs, the total import bill for fertilizer would be $1.67 billion a year. Yearly consumption of furnace oil would drop by 3 million tons, resulting in savings of $1.75 billion at the 2015 import price. The net savings in foreign exchange reserves would be $80 million a year.

Consumers would also benefit from the shift to imported fertilizer because the maximum local retail price exceeds the international benchmark price even after differences in transport costs are taken into account (Figure 5.6). Given the current domestic demand for urea and DAP, annual fertilizer costs would fall by $836 million. In addition, the government would save $163 million annually because it provides a subsidy of 50 Pakistani rupees (PRs) for each 50-kilogram bag of domestically produced DAP. Reducing support to domestic fertilizer production might also promote productivity of the domestic industry (Ali and others 2016).

Meanwhile, within the power sector the allocation of gas does not appear to follow a merit order. A 2013 World Bank study finds that, whereas the aging fleet of public plants (many of them more than 20 years old) are receiving gas, more modern and efficient private plants have often had to suspend production because of gas shortages (World Bank 2003). Because of lack of detailed data on plant availability and dispatch,

FIGURE 5.6 The domestic price of fertilizer in Pakistan is higher than the international price even after adding the transport cost

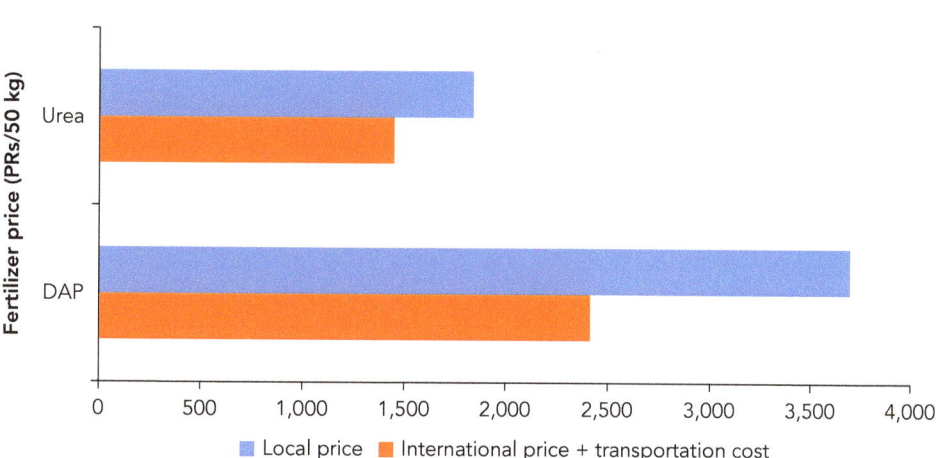

Source: Awan (2017).
Note: kg = kilogram; PRs = Pakistani rupees.

the cost of inefficient gas allocation in the power sector cannot be quantified. The simulation results therefore provide only a lower-bound estimate of the institutional cost of inefficient gas allocation.

INSTITUTIONAL: GAS LEAKAGE AND GAS THEFT

Adding to the problem of a low gas supply, Pakistan loses more than an eighth of gas during delivery. Two companies—Sui Southern Gas Company Limited (SSGC) and Sui Northern Gas Pipelines Limited (SNGPL)—manage most of the gas transmission and distribution in Pakistan. Both have performed poorly in managing distribution networks, leading to high levels of unaccounted for gas (UFG)—the difference between the volume of metered gas at the point of dispatch and the volume of gas sold to consumers. Gas losses have been rising sharply since 2010, peaking at a staggering 14.3 percent in fiscal 2015 (Figure 5.7). By comparison, UFG is typically 1–2 percent in Organisation for Economic Co-operation and Development countries because of both better management and the larger share of bulk consumption in the consumer mix.

Many factors have contributed to the high level of gas losses in Pakistan, but the main causes can be traced to the ways in which SSGC and SNGPL are regulated. No regulatory mechanism links their financial returns to their operational efficiency: Both companies are subject to rate-of-return regulation allowing a guaranteed return on their fixed assets of about 17 percent. Under this regulatory approach, tariff setting disproportionately rewards capital investment (Averch and Johnson 1962), encouraging firms to favor network expansion over pipeline maintenance.

FIGURE 5.7 A large share of gas is lost during transmission and distribution in Pakistan

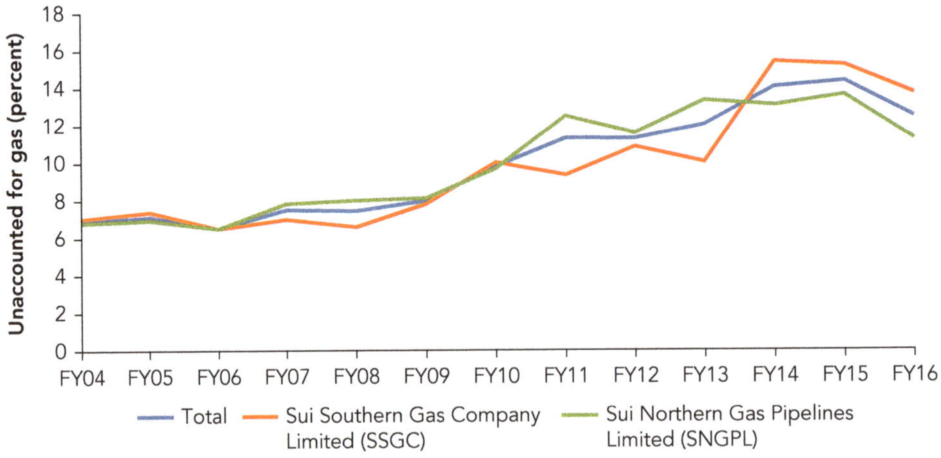

Source: KPMG (2017).
Note: FY = fiscal year.

Moreover, until 2002 the cost of UFG was fully passed on through gas prices, making consumers the ultimate losers from the inefficiencies of gas companies. In 2002 the Oil and Gas Regulatory Authority (OGRA), an independent regulator for the mid- and downstream oil and gas industry, set targets for gas companies to reduce UFG. It initially set the performance benchmark for UFG at 4 percent, but then raised it to 7 percent. Above this benchmark, firms cannot recover the cost of UFG through retail tariffs. This kind of performance-based regulation can strengthen incentives to reduce losses. But shifting performance standards can undermine its effectiveness by inducing firms to expect a chance to renegotiate targets. Indeed, in 2017 gas companies were lobbying for a pass-through of costs for up to 11 percent UFG.

The ownership of SSGC and SNGPL also plays a part in operational inefficiency. Although both are owned by a mix of public and private shareholders, the government is the largest shareholder, which can lead to interference to achieve politically motivated goals.

All of these factors have contributed to rapid expansion of gas networks and neglect of maintenance—and consequently to an increase in pipeline leakage and gas theft. Both SSGC and SNGPL have expanded their transmission networks rapidly over the past two decades. As a result, Pakistan now has one of the most extensive inland gas supply systems in the world. With a total length of roughly 158,000 kilometers by the end of fiscal 2016, it is long enough to circle the earth more than three times (Figure 5.8).

Not surprisingly, according to an OGRA-sponsored study, underground leakage from aging pipelines and poor maintenance have contributed significantly to UFG in Pakistan. The average leakage rate is 4.9 leaks per kilometer for SSGC and 2.2

FIGURE 5.8 Gas connections and the gas network have expanded rapidly in Pakistan

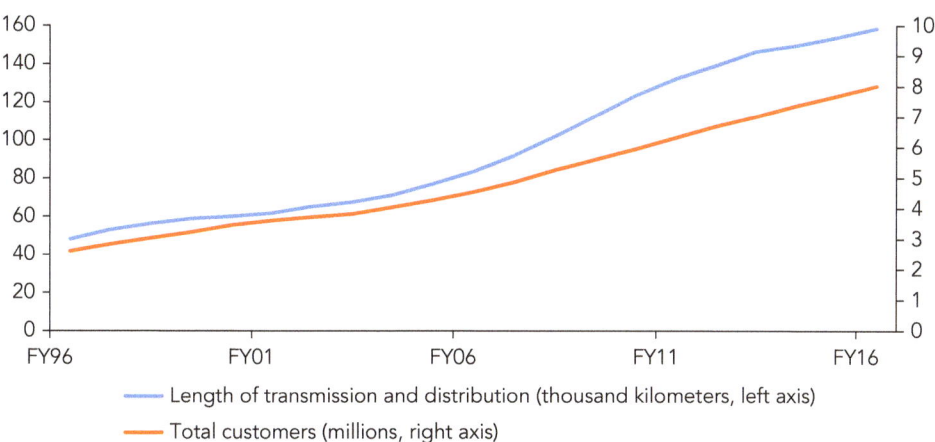

Length of transmission and distribution (thousand kilometers, left axis)

Total customers (millions, right axis)

Source: Ministry of Energy, Pakistan (2001, 2007, 2010, 2014, 2016).
Note: FY = fiscal year.

for SNGPL. By contrast, the average is 0.215 in Germany and 0.36 in the U.S. state of Massachusetts (KPMG 2017). Other contributing factors include gas theft and the shift of gas consumption from the bulk to retail sectors. Gas theft accounted for 16.5 percent of total UFG in 2015, including 14 percent by unregistered users illegally drawing gas from pipelines and 2.5 percent by registered users tampering with meters. The expansion of gas consumption in the retail sector increased overhead leakage (occurring because pipelines for residential connections are aboveground and exposed to the environment) and increased the incidence of theft.

Optimizing the allocation of gas and improving efficiency in gas delivery would shift the supply curve of gas to the right, so that at any given price level the new equilibrium supply would be higher. To estimate the institutional costs of both the inefficient allocation and inefficient delivery of gas, the analysis simulates this new gas supply curve and projects the new equilibrium quantity of gas supply, assuming that all gas used by the fertilizer sector is diverted to power generation and UFG is reduced to 4.5 percent, an

---

BOX 5.1 Estimating the price elasticity of supply and demand for gas in Pakistan

This analysis estimates the price elasticity of the domestic gas supply and the price elasticity of the demand for gas by the power sector in Pakistan, using income and price as the main determinants of supply and demand. The estimation is based on annual data for domestic gas production, the real weighted-average retail price for gas (deflated by the consumer price index), the real average gas price for the power sector, gas consumption of the power sector, and the real per capita GDP over 1995–2016. These data are obtained from various issues of the *Pakistan Economic Survey* and various editions of the *Pakistan Energy Yearbook*.

To address the potential endogeneity of price, the analysis instruments the contemporaneous gas price by the first two lags of the price of gas and the first two lags of the real wage index for mining and quarrying in the estimation of the supply curve (wage data are from the Pakistan Labor Force Survey). The analysis instruments the contemporaneous gas price by its first four lags in the estimation of the demand curve. It tests for the existence of long-run cointegration and estimates an error correction model to obtain both the short-run and long-run elasticities (Appendix A provides details on the methodology).

The estimated long-run supply and demand elasticities are 0.31 and –0.61, respectively. Thus a 1 percent increase in the domestic price of gas increases the supply of gas by 0.31 percent and reduces the power sector demand for gas by 0.61 percent. Short-term coefficients on prices are 0.06 for the supply curve and –0.09 for the demand curve. As expected, short-term elasticities are lower than the long-run estimates (in absolute terms). Analysis ignoring potential price endogeneity produces counterintuitive signs of price elasticities.

Several studies estimate the supply and demand elasticity for gas using rigorous econometric techniques. Khan (2015) finds that the price elasticity of the power sector demand for gas in Pakistan is –0.51 in the short run and –0.76 in the long run. And, according to Hausman and Kellogg (2015), in the United States the long-run price elasticity of gas supply is 0.81 and the long-run price elasticity of power sector demand for gas is –0.47. The estimated price elasticities in this analysis are generally lower than those suggested in the literature. With larger elasticities, the deadweight loss associated with power sector distortions would be higher.

initial target set by OGRA. With the increase in gas supply, consumers would benefit from reduced unmet gas demand and reduced expensive imports of LNG. On the basis of the estimated price elasticity of supply and demand, the consumer welfare gain is estimated at $720 million a year (Box 5.1). Producers would benefit from higher profits. The producer surplus change is estimated at $150 million a year. More efficient gas allocation would reduce the cost of oil imports (after deducting the increase in fertilizer imports) by $80 million a year and subsidies for domestic fertilizer by $163 million a year. The total welfare loss from upstream institutional distortion is therefore estimated at $1.12 billion (0.41 percent of gross domestic product, GDP) a year in fiscal 2015.

REGULATORY: UNDERPRICED GAS

Natural gas prices in Pakistan are regulated at both the wholesale and retail level. The wellhead prices are linked to international oil prices within a prescribed band (with a floor of $30 per barrel and a ceiling of $110 per barrel). Each gas field also has a separate price formula, depending on the timing of its initial exploration contract, its location, or its gas composition. For SSGC and SNGPL, which function as monopoly buyers and sellers of gas, the weighted-average purchase cost for wellhead gas is estimated at $2.63 per million British thermal units (MMBtu) in fiscal 2017. By comparison, the international market price is $8.30 per MMBtu.

At the retail level, there is a two-tier gas market for domestically produced gas and imported LNG. The price for domestic gas is set by OGRA at a level that allows SSGC and SNGPL to pass through their costs and earn an agreed-on rate of return on their investment. Although the retail price is higher on average than the supply cost, the weighted-average retail price and the retail price for the power sector are consistently lower than the international benchmark price, the correct measure of the opportunity cost of gas supply (Figure 5.9). In addition, cross-subsidies exist between consumer categories, leading to extremely low tariffs for the residential and fertilizer sectors and high demand for low-value uses of gas (Figure 5.10). Imported LNG is separated from domestic gas streams and allocated to specific customers at a separately regulated price. ORGA is mandated to determine the price of LNG on a monthly basis. The average LNG price was $8.68 per MMBtu in fiscal 2016. This price of imported gas was 1.8 times the price of domestic gas for power generation and 2.2 times the weighted-average domestic gas price in fiscal 2016.

Gas subsidies can also be evaluated against the cost of the cheapest alternative fuel, furnace oil. Although the real weighted-average gas price for power generation has been hovering at around PRs 160 per gigajoule (GJ) since fiscal 2002, the real price for furnace oil increased at a compound annual growth rate of 6 percent a year during fiscal 2002 and 2014 (Figure 5.11). Falling international oil prices helped reduce the gap between prices for gas and furnace oil during fiscal 2015 and 2016. But, when oil prices rose in fiscal 2017, the gap increased again; the price of furnace oil rose to 2.1 times the price of gas for power generation.

FIGURE 5.9 The price of domestic gas is substantially below the price of imported LNG in Pakistan

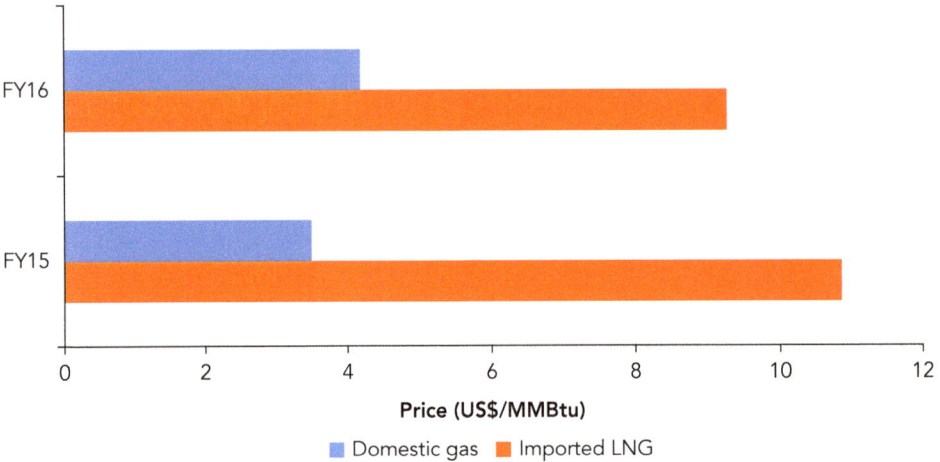

Source: Price of imported liquefied natural gas (LNG): OGRA (2016, 2017); weighted-average price of domestic gas: Ministry of Energy, Pakistan (2016, 2017) and Ministry of Finance, Pakistan (2017).
Note: FY = fiscal year; MMBtu = million British thermal units.

FIGURE 5.10 The gas price is most heavily subsidized for residential users and the fertilizer sector in Pakistan

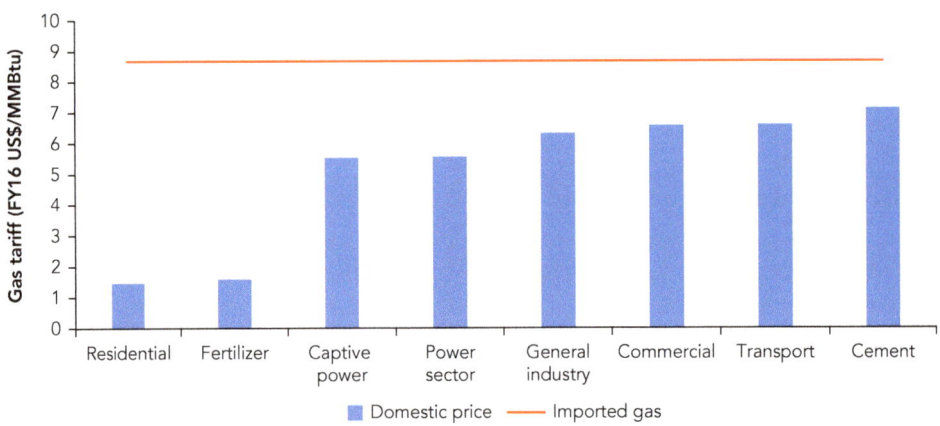

Source: Ministry of Energy, Pakistan (2017); OGRA (2017); exchange rate: World Bank, World Development Indicators (database).
Note: FY = fiscal year; MMBtu = million British thermal units.

Low tariffs are weak incentives for gas exploration and production. Lack of competition and an inability to directly access the end-user market further reduce incentives for producers to invest and innovate. Although SSGC and SNGPL are the government-designated buyers, gas companies are allowed to sell 10 percent of their output to third parties (other than residential and commercial consumers) at negotiated prices. They must obtain prior government consent, however, and pay a gas windfall tax of 40 percent of the difference between the regulated and negotiated price. These requirements effectively prohibit direct sales to third parties.

Low retail prices also encourage excessive gas demand. In addition, through financial transfers between SSGC and SNGPL, the government implements a uniform gas price for each consumer category across the country regardless of the distance, volume, or cost of supply. This policy further insulates consumers from market conditions and makes them less sensitive to cost changes.

The economic waste from gas underpricing consists of (1) the welfare losses of consumers who are willing to pay more for gas than the international market price but remain unserved under the current pricing regime and (2) the losses of producers who would profit from higher prices and expanded production. Analysis to estimate the cost of inefficient gas pricing uses estimated supply and demand equations to predict a new market equilibrium, assuming that the price of domestic gas is raised to the price of imported LNG (Box 5.1). Raising the price to this level would boost domestic gas production by 29 percent and reduce the domestic demand for gas to

FIGURE 5.11 Gas for power generation is cheaper than the cost of replacement fuels in Pakistan

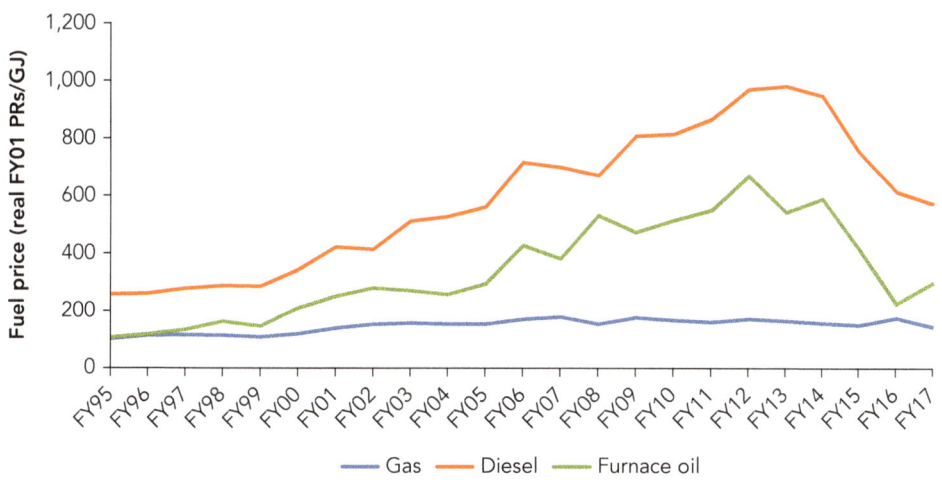

Source: Ministry of Energy, Pakistan (2016, 2017); OGRA (2017); Pakistan State Oil (2018); consumer price index: Ministry of Finance, Pakistan (2003, 2015, 2017).
Note: FY = fiscal year; GJ = gigajoule; PRs = Pakistani rupees.

the extent that there would be no gas shortage in the domestic market. Although consumers would face higher gas prices, their losses would be more than offset by gains in producer surplus, estimated at $2.58 billion a year. The net welfare loss from departing from this efficient pricing is estimated at $348 million a year (0.13 percent of GDP in fiscal 2015).

SOCIAL: EMISSIONS FROM THE USE OF GAS AND OIL

Pakistan has among the highest levels of air pollution in the world, and fossil fuel–based power generation has been identified as an important source (Sanchez-Triana and others 2014). Burning gas and oil releases volatile organic compounds, a group of chemicals that contribute to the formation of ground-level ozone (urban smog), exposure to which is linked to a wide range of health effects. Ozone levels in Pakistan have been rising. Indeed, among the world's most populous countries, Pakistan had the third-largest increase in seasonal average population-weighted concentrations of ozone over the past 25 years (Health Effects Institute 2017). In 2015 exposure to ozone contributed to 5,000 deaths in Pakistan, up 213 percent from 1990.

Gas- and oil-based power generation is also a major source of greenhouse gas emissions. Gas is cleaner than coal when it is burned. When it is leaked, however, it releases methane, an extremely potent greenhouse gas. Over the course of a century, methane will trap 28–34 times as much heat in the atmosphere as an equivalent amount of carbon dioxide. One study suggests that a gas plant will cause more climate damage than a coal plant if it leaks just 3 percent of its gas into the air before combustion (Alvarez and others 2012). Given the high level of UFG, Pakistan could easily exceed this threshold.

Improving the efficiency of gas allocation could help address some pollution issues. Furnace oil, with its higher content of sulfur, methane, and metals, is a bigger source of air pollution than gas. Reallocating gas from fertilizer production to the power sector could reduce oil consumption by 3 million tons a year, leading to a corresponding annual reduction in carbon dioxide emissions of 1.84 million tons. On the basis of the assumption that the shadow price of carbon dioxide emissions is $40 per ton, the external cost from the extra emissions is $70 million a year.

Imposing an environmental tax on gas consumption could be another way to address the external costs of emissions. A global study estimates the marginal health and climate damages from gas combustion in Pakistan at $0.17 and $2.41 per GJ, respectively (Parry and others 2014). Levying an environmental tax equal to the sum of these marginal social damages would not, however, change the current level of gas consumption in Pakistan because demand would consistently exceed supply with or without environmental pricing.

The total cost of social distortion in the upstream gas sector—the environmental cost from excessive use of oil caused by inefficient gas allocation—is estimated at $70 million a year. Because of data limitations, this estimate does not include the health costs of increased oil consumption and the environmental and health costs of gas leakage. This result is therefore a lower-bound estimate of the actual cost.

Core

Historically, two integrated public utilities dominated the core power sector in Pakistan. The Water and Power Development Authority (WAPDA) served most of the country; the Karachi Electric Supply Company (KESC) served the city of Karachi and surrounding areas. Both functioned satisfactorily until the mid-1980s, when public spending constraints led to inadequate investment in generation capacity and in the transmission and distribution infrastructure. By the beginning of the 1990s, power supply could not keep pace with demand.

In response to shortages, Pakistan initiated a power sector reform program in the early 1990s. It began allowing the entry of independent power producers (IPPs) in 1994; formed an independent regulator, the National Electric Power Regulatory Authority (NEPRA), in 1997; shifted from a monopoly to a single-buyer model by unbundling WAPDA into separate generation, transmission, and distribution companies in 1998; and privatized KESC in 2005. The latest national power policy, issued in 2013, envisions the development of a competitive wholesale electricity market. As a first step, the government separated the market settlement function from the national transmission company.

Power sector reform followed a textbook model. But more than 20 years after the reform was launched, Pakistan still suffers from prolonged load shedding and expensive power generation. Persistent institutional and regulatory shortcomings are among the factors contributing to the sector's weak performance.

INSTITUTIONAL: INEFFICIENT GOVERNMENT-OWNED POWER PLANTS

Since the government first began permitting private power generation projects in 1994, foreign and domestic investors have added substantial capacity to the national grid. By fiscal 2016, the private sector accounted for 45 percent of generation capacity and 53 percent of electricity output (Figure 5.12). Almost all private power plants are thermal, with 64 percent using gas, 33 percent furnace oil, and the rest coal, diesel, and hydro. Public plants own most of the hydroelectric and all of the nuclear units. Although the thermal capacity of public and private plants increased in equal proportion, output from IPPs rose from 30 percent of total thermal electricity in 1997 to 66 percent in fiscal 2016.

Regardless of ownership type, plants face only weak incentives to improve efficiency because of a compensation scheme that provides no direct link between operating efficiency and economic return. Power purchase agreements offer compensation on a

FIGURE 5.12 The public sector still accounts for a large share of electricity
generation in Pakistan

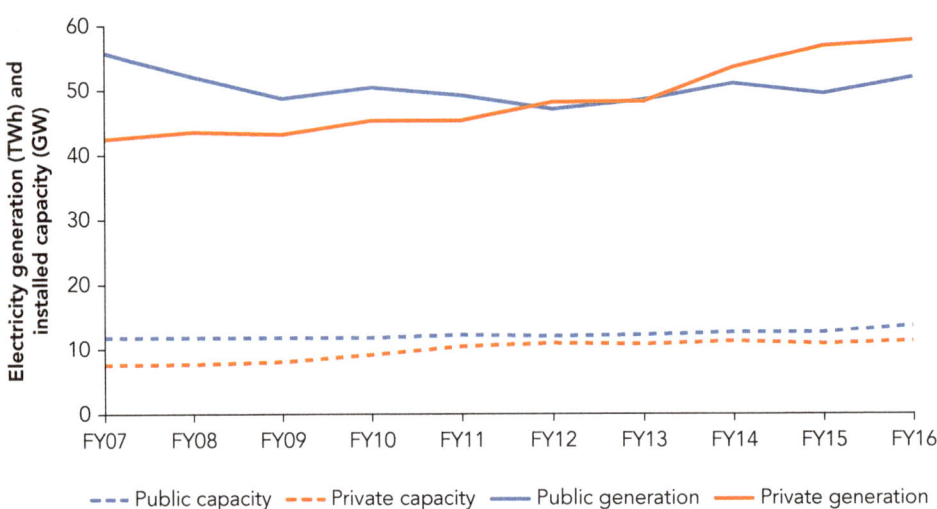

Source: Ministry of Water and Power, Pakistan (2006, 2010, 2014, 2016); Hydel data, NEPRA (2006–14).
Note: FY = fiscal year; GW = gigawatt; TWh = terawatt-hour.

cost-plus basis and a guaranteed return on equity. Fuel costs and variable operating and maintenance costs automatically pass through to generation tariffs, and plants receive a fixed capacity payment under long-term "take or pay" contracts. For public sector plants, the lack of a profit motive and of transparency and accountability in operations further undermines efficiency incentives.

To gauge the efficiency penalty associated with being a public thermal power plant, this analysis relies on plant-level input and output data from two main sources. NEPRA's annual *State of Industry Report* (2006–15 editions, the most recent editions available at the time of analysis) provides detailed information on energy consumption, electricity output, and operating and maintenance costs for 66 thermal power stations during fiscal 2006–15. The *Pakistan Energy Yearbook*, published by the Ministry of Petroleum and Natural Resources and the Hydrocarbon Development Institute of Pakistan, presents information on electricity output but not fuel inputs. The analysis also uses data on plant characteristics such as vintage, size, and technology collected from the generation licenses of power plants, as published by NEPRA, the Platts database, and Power System Statistics.

Each *State of Industry Report* publishes both historical and current data. The historical data are often revised in later reports, and the revisions appear to be especially significant after 2012. In addition, there is discrepancy between generation data reported by NEPRA and the Ministry of Petroleum and Natural Resources and the Hydrocarbon Development Institute of Pakistan. For a robustness check, three data sets were compiled: one based on data most recently published by NEPRA, one based on data first published in the corresponding year by NEPRA, and one based on generation data from the Ministry of Water and Power and the Hydrocarbon Development Institute of Pakistan and the latest input data from NEPRA. The results reported here are based on the first data set, but the main qualitative conclusions are robust to alternative data sets.

A simple descriptive analysis of the data reveals a large efficiency gap between public and private power stations (Figure 5.13). During fiscal 2006–15, among plants using gas, the median fuel efficiency (ratio of electricity output to heat input) was 41 percent

FIGURE 5.13 Public power plants are less efficient than private power plants in Pakistan

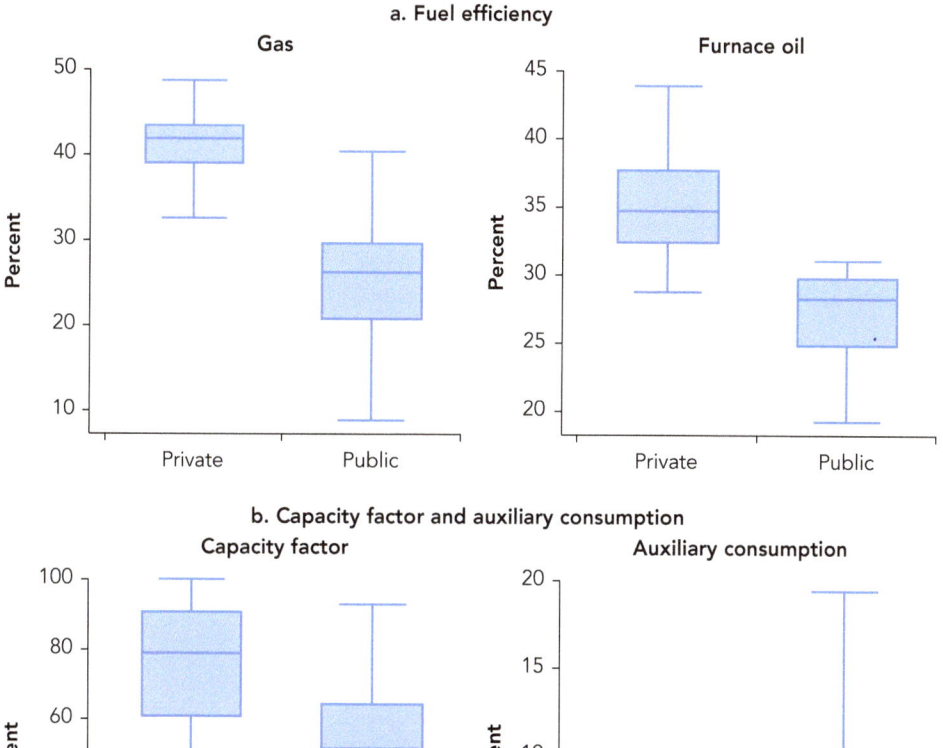

Source: Based on plant-level data, NEPRA, *State of Industry Report* (2006–15).
Note: Fuel efficiency is the ratio of electricity output to heat input. The graphs compare the median values of the plant's fuel efficiency, capacity factor, and auxiliary consumption without controlling for other confounding factors.

for private plants and only 26 percent for public ones. Among plants using furnace oil, the median fuel efficiency was 35 percent for private plants and 28 percent for public plants. In addition, although private plants use only 2.9 percent of the electricity they produce to run auxiliary equipment such as boilers and fans, public power stations consume 6.7 percent. The capacity factor was also higher for private plants than for public plants, with a median value of 79 percent and 52 percent, respectively.

Private and public plants also differ systematically in several physical and technological attributes that could contribute to their differences in efficiency. Private plants are newer and much more likely to use combined-cycle units rather than internal combustion or steam turbines. The disparity in efficiency persists even when the analysis controls for these differences, however. The analysis estimates a fuel input demand function using a plant's total generation, size, age, technology, and ownership type as explanatory variables. It also controls for yearly shocks common to all plants. The results show that, for an equivalent amount of electricity generation, public plants use 17–28 percent more fuel than private plants, depending on the type of fuel (Figure 5.14). The efficiency gap is larger for power plants using furnace oil.

Another factor affecting the fuel efficiency of a thermal power plant is how often it is ramped up and down and how often it operates below capacity. Shutdowns and

FIGURE 5.14 Public power plants in Pakistan are less efficient than private plants even after controlling for their characteristics

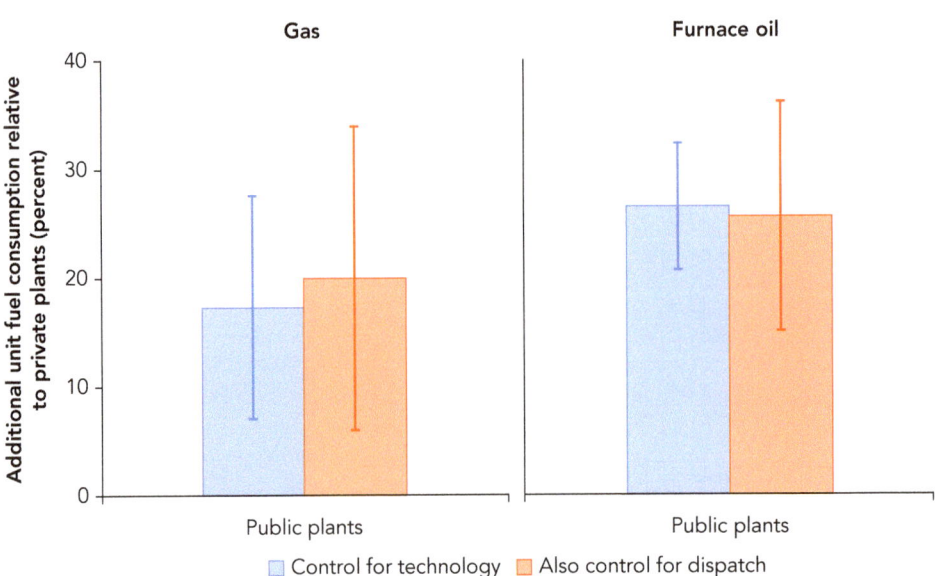

Source: Based on plant-level data, NEPRA, State of Industry Report (2006–15).
Note: Fuel intensity measures fuel input (in gigajoules) per unit of electricity output. The orange bars show estimated coefficients from regression analysis controlling for dispatch through the control of provincewide electricity output. The blue bars show estimated coefficients from regression analysis not controlling for dispatch. Bars denote point estimates and lines denote 95 percent confidence interval.

underutilization may reflect inefficiency, but they also can result from gas shortages or from dispatch not always following merit order (because of favoritism or technical constraints such as transmission congestion). If there are differences in the treatment of public and private plants in gas access, plant dispatch, or both, failing to control for these differences can lead to biased conclusions or interpretations.

The analysis cannot directly control for a plant's utilization rate because it is determined at least in part by the plant's efficiency. To address this simultaneity concern, the analysis follows Fabrizio, Rose and Wolfram (2007) by using fluctuation in provincewide electricity demand as a source of exogenous variation in plants' dispatch. Regardless of its ownership type, a power plant is more likely to be dispatched and to run at higher capacity when there is a provincewide surge in demand. But provincewide demand is unlikely to be correlated with an individual plant's efficiency.

After controlling for potential differences in plants' utilization rates, the difference in efficiency between the public and private sectors increases for gas-based power plants, but it remains unchanged for furnace oil–based plants (Figure 5.14). The analysis shows that, compared with a private plant, a public gas plant uses 20 percent more gas for each unit of electricity produced. This finding is consistent with anecdotal evidence that public plants had privileged access to gas; without it, they would have performed even worse than they do.

In addition to comparing single-factor productivity (fuel intensity) between different type of plants, the analysis examines the correlation between ownership type and total factor productivity (TFP). Factor prices could affect single-factor productivity. For example, firms facing lower fuel prices might choose to use fuel more intensively relative to other inputs. By contrast, TFP is measured as the residual of a production function that controls for all observable inputs—and it is thus invariant to the intensity of use of factor inputs. Using the same level of inputs, generators with higher TFP will produce more electricity than generators with lower TFP.

Employing plant-level panel data from fiscal 2006–15, the analysis uses stochastic frontier analysis to estimate the production function of power stations in Pakistan. This approach has the advantage of allowing stochastic noise in the data and separating random productivity shocks and systematic technical inefficiency in the estimation (see Appendix B). Within a stochastic frontier setting, the distribution of inefficiency is considered to be a function of the ownership type of plants. In addition to fuel, capital expenses (proxied by capacity), and operating and maintenance costs, the analysis controls for a time trend in the production function.

The results show that private plants are much more productive than public plants (Figure 5.15). The average technical efficiency score (the ratio of actual output to maximum feasible output) during the sample period is 0.82 for private plants and 0.55 for public plants. Across the entire distribution of power plants, a randomly selected public plant always has a lower technical efficiency score than a randomly selected private plant.

Because the analysis controls for plants' exogenous physical and operational characteristics and their observable inputs, the remaining differences in operating efficiency

FIGURE 5.15 Measured by total factor productivity, public power plants are less efficient than private plants in Pakistan

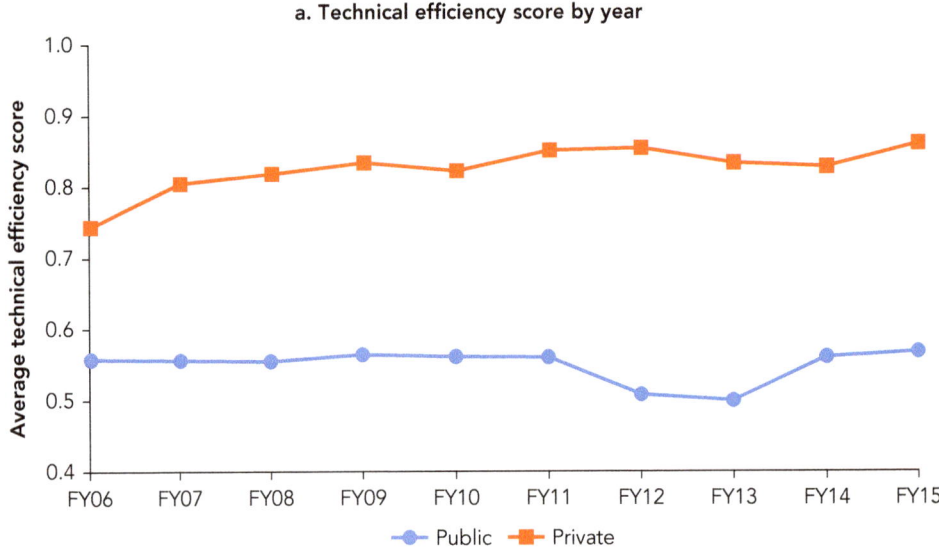

a. Technical efficiency score by year

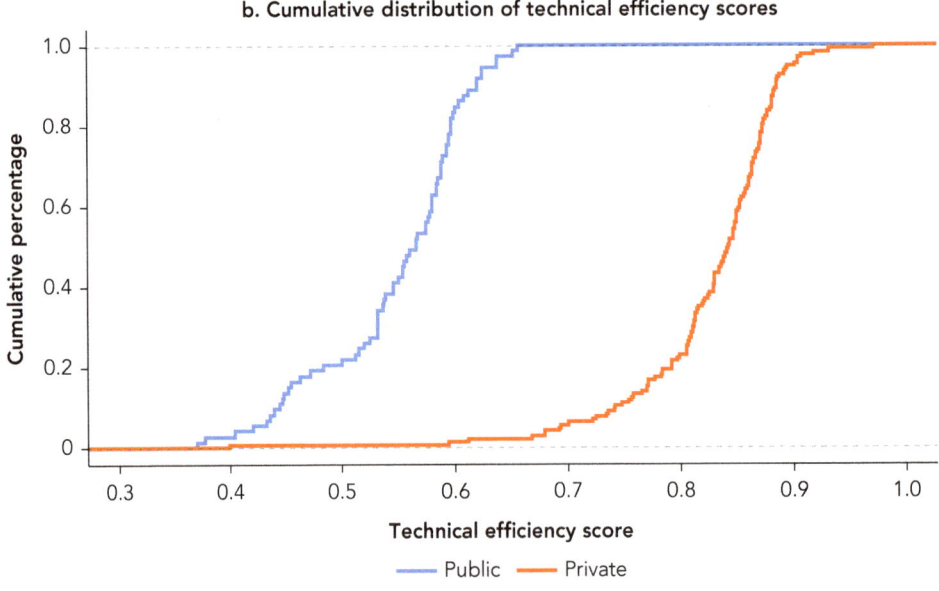

b. Cumulative distribution of technical efficiency scores

Source: Based on data from NEPRA, *State of Industry Report* (2006–15).
Note: A technical efficiency score measures the ratio of actual output and maximum feasible output.
FY = fiscal year.

between private and public plants are likely to be correlated with innate differences in these two types of plants. Some of these differences may be explained by the type of power purchase agreements signed by private plants. Anecdotal evidence suggests that some private plants are allowed to be dispatched only at the optimal load factor to maximize fuel efficiency while public plants would need to take up the slack. However, because power purchase agreements are mostly confidential, the extent of such preferential treatment is unknown.

To the extent that the difference in efficiency between public and private plants also reflects differences in the quality of their management, institutional reforms aimed at improving the incentive for public utilities to improve managerial performance could yield large gains. To quantify the potential gain from addressing institutional shortcomings in the generation sector, a simulation analysis is conducted to estimate how much additional electricity would be produced if gas-based public plants matched the managerial performance of private plants—that is, reduced their fuel intensity by 20 percent. The analysis simulates daily load shedding and generation profiles in Pakistan using the daily dispatch reports of the National Transmission and Dispatch Company (NTDC) and plant-level input and output data from March 2014 to April 2015 (the period for which the daily reports are available). The simulation assumes that each plant uses its own fuel savings from efficiency improvements up to its maximum capacity. The additional production, subject to a 19 percent transmission and distribution loss, is used to reduce the unserved demand for electricity. The simulation results show that the efficiency improvement alone could reduce the unserved energy demand by 25 percent (Figure 5.16).

FIGURE 5.16 Improving generation efficiency would reduce power shortages in Pakistan

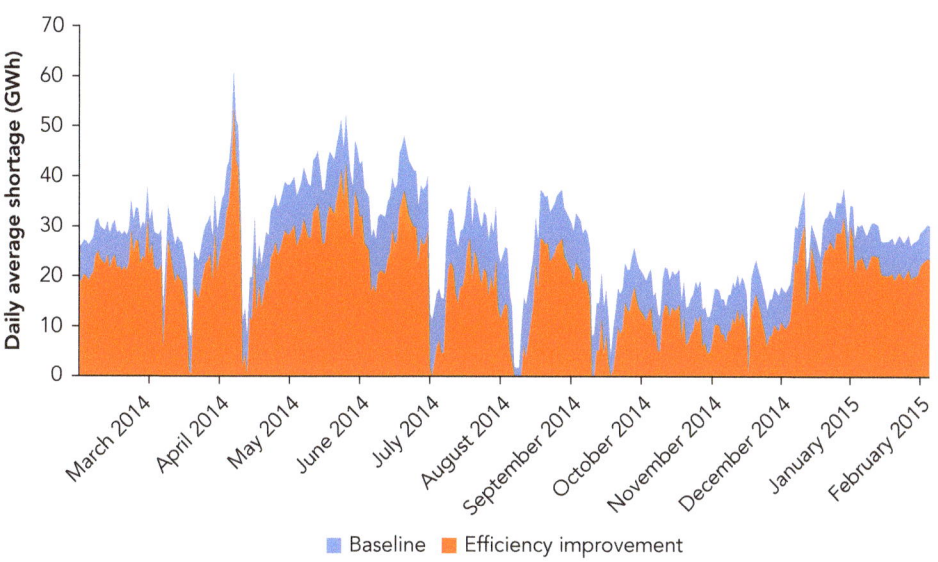

Source: Based on daily generation reports, National Transmission and Dispatch Company, and NEPRA (2006–15).
Note: GWh = gigawatt-hour.

Improving fuel efficiency would shift the electricity supply curve to the right. To estimate the welfare gain from removing institutional distortions in generation, the analysis first estimates the price elasticities of demand and supply for electricity using historical data on energy price and quantity in Pakistan (Box 5.2). It then simulates the new supply survey on the basis of the estimated output increase described earlier. Assuming subsidies remain at 7 percent of the cost of supply (for more details, see the section on underpriced electricity), the analysis projects the new equilibrium quantity and calculates the corresponding changes in consumer and producer surplus. Results from the analysis show that the lower price and greater supply of electricity lead to $486 million a year in benefits to consumers and $622 million a year in benefits to producers. However, government expenditures on subsidies would also increase following output expansion. The net welfare gain from removing institutional distortions in generation is thus estimated at $966 million (about 0.4 percent of GDP) a year in fiscal 2015.

Lack of efficiency is a problem not just among public plants; even private generators operate below the production frontier. Experience in the United States suggests

BOX 5.2 **Estimating the price elasticity of supply of and demand for electricity in Pakistan**

This analysis estimates the price elasticity of electricity supply and demand in Pakistan, using income and price as the main determinants of supply and demand. The estimation is based on annual data for electricity generation, the estimated shortage in electricity supply, the real weighted-average electricity price index, and real per capita GDP over the period 1994–2013. These data are obtained from various issues of *Power System Statistics*, published annually by the NTDC, and *Electricity Demand Forecast Based on Multiple Regression*, published by NTDC in 2014.

To address the potential endogeneity of price, the analysis instruments the contemporaneous electricity price by the first five lags of electricity price in the estimation of the supply curve. It instruments the contemporaneous electricity price by the first lag of the electricity price and the first lag of the furnace oil price in the estimation of the demand curve. The analysis tests for the existence of long-run cointegration and estimates an error correction model to obtain both short- and long-run elasticities (Appendix A provides details on the methodology).

The estimated long-run supply and demand elasticities are 0.41 and –0.45, respectively, meaning that a 1 percent rise in the price of electricity increases the supply of electricity by 0.41 percent and reduces the demand for electricity by 0.45 percent.

Several studies estimate the price elasticity of electricity demand in Pakistan. Using firm-level data for 2002–06, Chaudhry (2016) finds that it ranges from –0.31 in the electronics sector to –0.81 in the textile sector. Using household-level data, Nasir, Tariq, and Ankasha (2008) find short- and long-run elasticities of –0.63 and –0.77, respectively, for the residential sector. The NTDC estimates the price elasticity of electricity demand for various sectors using data for 1970–2006. It finds price elasticities of –0.14 for the residential sector, –0.28 for the agricultural sector, and –0.55 for the commercial sector. Elasticities estimated in this analysis are in line with estimates reported in the literature.

that replacing rate-of-return regulation with market pricing (such as through a competitive wholesale market) could yield further efficiency gains. In anticipation of greater competition in generation, private plants in U.S. states undergoing market deregulation reduced their operating expenses by up to 5 percent relative to private plants in states that continued to use rate-of-return pricing (Fabrizio, Rose, and Wolfram 2007; Zhang 2007). The results presented here are therefore a lower-bound estimate of the institutional cost in the generation sector.

INSTITUTIONAL: HIGH LOSSES OF DISTRIBUTION UTILITIES

There are 10 distribution companies in Pakistan. Despite the government's ambitious plans to privatize distribution and generation companies, the Ministry of Energy controls and manages all 10. Only in Karachi is distribution privately provided, by K-Electric, a privately owned, vertically integrated utility that controls generation, transmission, and distribution.

Distribution is hugely inefficient in Pakistan, but there is strikingly wide variation in performance across distribution companies. In 2016 almost a fifth of the electricity generated was lost in the network as a result of both technical reasons and electricity theft (Figure 5.17). Three distribution companies (IESCO, GEPCO, and FESCO) managed to keep losses around the international standard of 10 percent, however, even as two others (PESCO and SEPCO) incurred losses above 35 percent.

FIGURE 5.17 Transmission and distribution losses are high in Pakistan

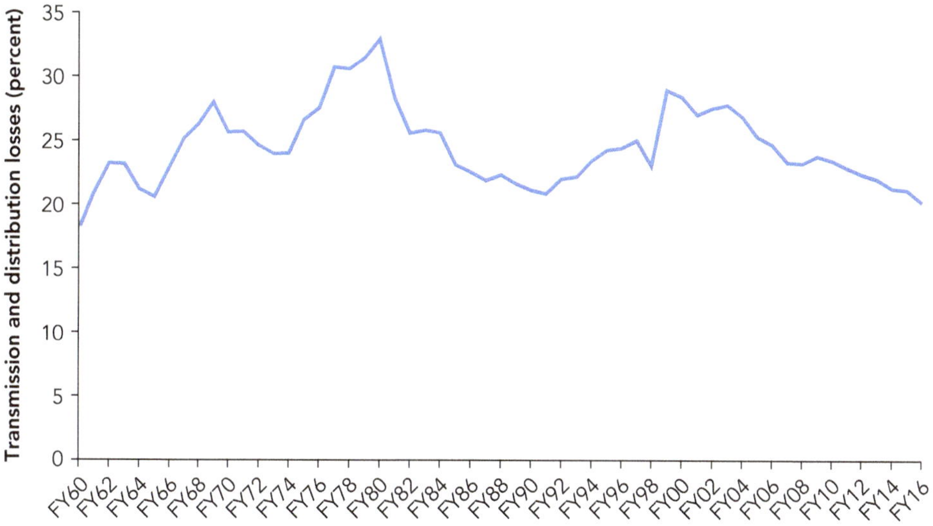

Source: NTDC (2016).
Note: FY = fiscal year.

Other performance indicators reveal equally large differences in efficiency (Figure 5.18). The revenue collection rates of distribution companies range from 32 percent to more than 100 percent, and the frequency of power supply interruptions (excluding load shedding) during fiscal 2015 ranged from 36,879 times for TESCO to 890 million times for MEPCO.

The dispersion in efficiency across distribution companies persists even after controlling for potential external drivers of productivity. The variation in performance could be linked to differences in managerial performance, but it could also reflect factors beyond the control of distribution companies. For example, electricity losses in transmission and distribution might be greater in places with adverse weather conditions, lower population density, higher peak-load demand, higher levels of lawlessness, or a larger share of residential consumers.

To separate the effects of managerial performance, the analysis uses a stochastic frontier model to estimate a production function of electricity distribution using

FIGURE 5.18 Operational performance varies widely across distribution utilities in Pakistan

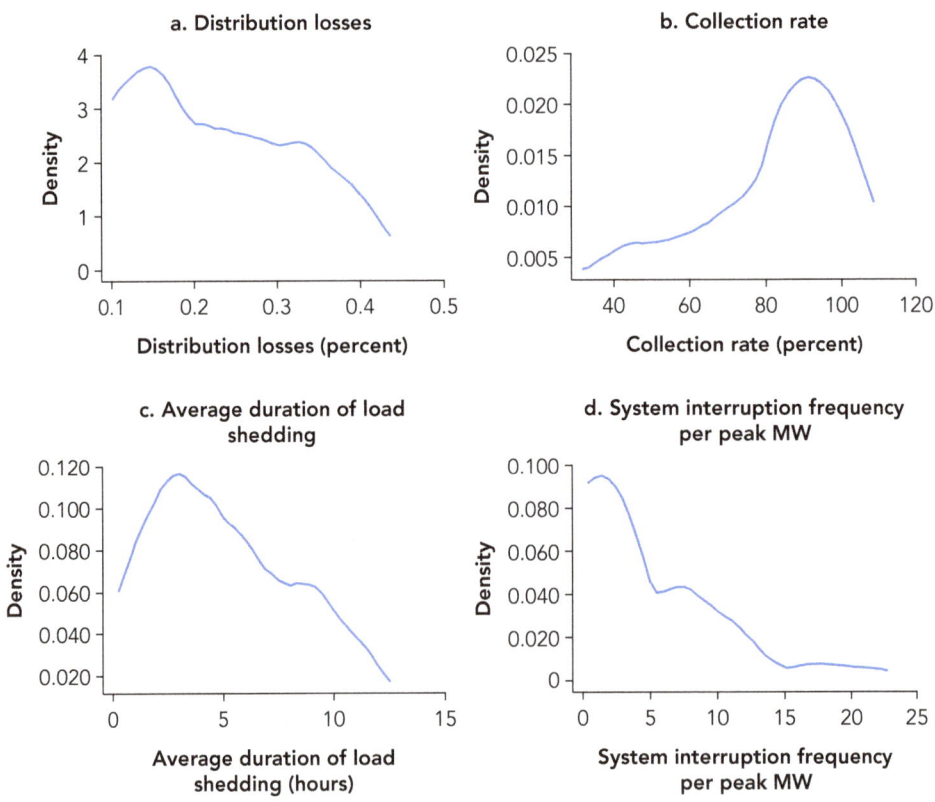

Source: NEPRA, *State of Industry Report* (2006–15).
Note: MW = megawatt.

annual panel data for fiscal 2011–15. The dependent variable is total electricity sold. The explanatory variables include the volume of electricity generated; the volume of electricity purchased; the location of the distribution company's service areas; the consumer mix (shares of residential, commercial, industrial, and agricultural consumers as well as public lighting, bulk supply, and others); and peak-load demand.

The results reveal a wide dispersion of efficiency across companies. Given the same electricity inputs, the most efficient distribution company sells 20 percent more electricity than the least efficient one, all else being equal. The average technical efficiency score in the sector is 0.87, meaning that the actual sale of power is 87 percent of the maximum feasible (Figure 5.19). On average, 13 percent of electricity is thus lost to inefficiency in transmission and distribution.

The large variation in performance across distribution companies suggests great potential to reduce losses through changes in managerial practices. Large differences in efficiency after controlling for many external factors can be interpreted as driven primarily by innate differences such as managers' skills or the quality of their practices.

Addressing institutional shortcomings—through privatization or incentive-based regulation, for example—could improve managerial performance. To estimate its potential gain (or the potential loss of institutional distortions), the analysis considers

FIGURE 5.19 Technical efficiency scores varies widely across distribution utilities in Pakistan

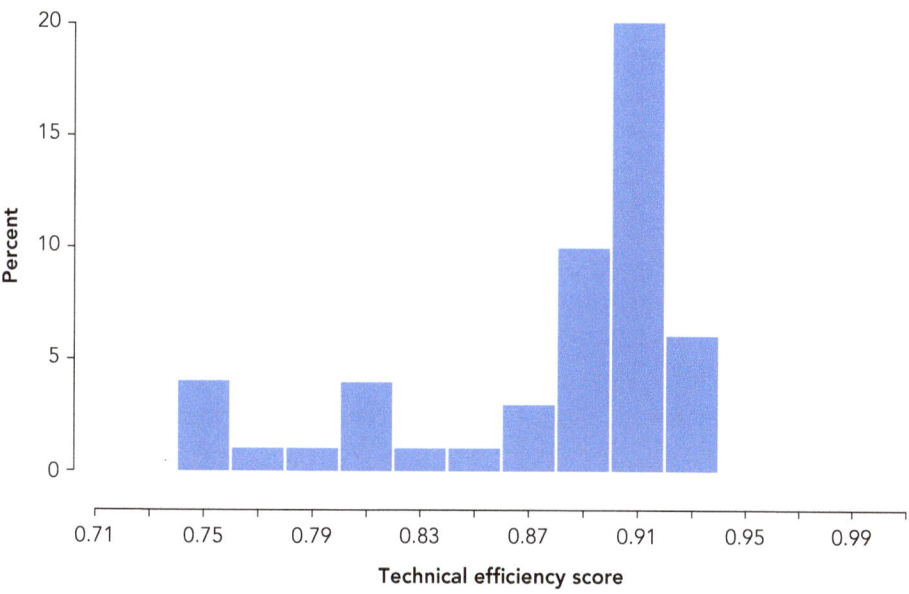

Source: Based on utility-level data, NEPRA, *State of Industry Report* (2011–15).

a counterfactual in which all distribution companies achieve the same technical efficiency score as the best performer among them. Such a scenario could reduce annual
aggregate transmission and distribution losses by 5.7 TWh. Reducing network losses
would shift the supply curve of electricity to the right. The analysis predicts the new
supply curve and new equilibrium quantity on the basis of the potential increase in
supply and the estimated price elasticities for supply and demand (see Box 5.2). The
corresponding consumer and producer surplus changes net the increase in subsidies
payment are estimated at $860 million (0.32 percent of GDP) in fiscal 2015.

INSTITUTIONAL: UNDERINVESTMENT IN TRANSMISSION

Beyond the high electricity losses incurred by distribution companies, substantial
line losses result from poor infrastructure and greater constraints in the transmission
system. The transmission of electricity from generation plants to distribution companies is controlled largely by the NTDC, a public company responsible for transmission
in all parts of the country except those served by K-Electric. In recent years, investment in the state-run transmission system has failed to keep pace with demand, causing
overloading and frequent tripping of grids. NEPRA estimates that 60 percent of the
550/220 kilovolt (kV) and 220/132 kV transformers in NTDC's system were overloaded
in 2015. Power plants with a total generating capacity of 1,600 MW faced evacuation
problems because of transmission constraints in the system (NEPRA 2015).

Distribution companies also face constraints in their own transmission networks.
NEPRA estimates that in 2015 more than half of the transformers in the distribution
network (excluding K-Electric) were overloaded. With transmission capacity barely able
to absorb the current generation capacity, the problems will only increase because of
the expected growth in demand and generation.

The financial constraints of state-owned utilities are among the main reasons for the
lack of investment in the transmission network. The government's intention to keep
most transmission under state ownership has limited the scope for private investment,
although in fiscal 2016 NEPRA took a step toward opening the transmission segment to
private construction and ownership of lines by approving a special-purpose transmission license for a private investor (NEPRA 2016).

Another important factor constraining transmission is transmission pricing,
which provides little signaling on where investment should be directed. In general,
when there is congestion in the network, the cost of transferring power differs in different nodes of the network, but the pricing mechanism does not reflect differences
in the value of electricity at different locations. Instead, a fixed fee is charged for
the use of transmission regardless of the distance, location, or related physical constraints of the system. Pricing therefore provides little incentive for investors to build
more transmission capacity in congested areas. As NEPRA notes, the NTDC could
not "prioritize its investment for improving those grids where tripping occurred"
(NEPRA 2015). Uniform transmission pricing will become an even bigger problem

as the transmission segment moves toward greater private sector participation and a more decentralized investment pattern.

NEPRA estimates that transmission constraints in the state-run system and those of the distribution companies accounted for 29 percent of the electricity shortfall in fiscal 2015—1,219 MW, down slightly from 1,249 MW in fiscal 2014. Removing transmission constraints would therefore increase the power supply by about 8.7 percent in that year. On the basis of the potential increase in electricity supply, the analysis estimates the net welfare loss from transmission constraints at $1.1 billion (0.41 percent of GDP) in fiscal 2015.

REGULATORY: UNDERPRICED ELECTRICITY

High transmission and distribution losses are compounded by the underpricing of electricity. NEPRA determines the retail electricity tariff for each distribution company on the basis of its cost of delivering electricity to consumers, including its power purchase cost, its targeted transmission and distribution cost, and a guaranteed return on its capital investments. But the actual consumer tariff—the government-notified tariff—is uniform across the country and is on average substantially lower. The difference, known as the tariff differential subsidy, is payable by the government. In fiscal 2012, this subsidy averaged 26 percent of the NEPRA-recommended tariff (Figure 5.20). In the same year, the federal budget allocation for the subsidies to the distribution sector amounted to 1.2 percent of GDP.

The average retail tariff has been substantially increased since adoption of the 2013 national power policy. In addition, taking advantage of falling fuel costs, in November 2014 the government introduced a tariff rationalization surcharge allowing a distribution company to collect the entire uniform national tariff from consumers whenever its cost-recovery tariff determined by NEPRA is lower than that tariff. Funds collected through this surcharge are used to cross-subsidize consumers served by the higher-cost distribution companies.

Direct budgetary transfers to the power sector have declined considerably since fiscal 2013, but they still amounted to $2.15 billion (0.8 percent of GDP) in fiscal 2015. In addition to explicit subsidies, there are implicit subsidies through toleration of electricity theft and nonpayment by both government and private entities. Poor recovery of electricity bills has led to high commercial losses for distribution companies that are often not recoverable through NEPRA-approved tariffs.

The underpricing of electricity and the failure to collect electricity dues have contributed to a vicious "circular debt." Because governments do not pay subsidies in a timely manner and customers do not fully pay their electricity bills, distribution companies cannot pay electricity generators. As a result, they cannot pay fuel suppliers, which then cut off their fuel supply, leading to idled generation capacity and electricity shortages. Arrears accumulated along the entire chain of supply reached a staggering level of almost PRs 872 billion (about 4 percent of GDP) by the end of fiscal 2012 (USAID 2013). When a new government took office in 2013, it immediately paid off PRs

FIGURE 5.20 Electricity prices are subsidized in Pakistan

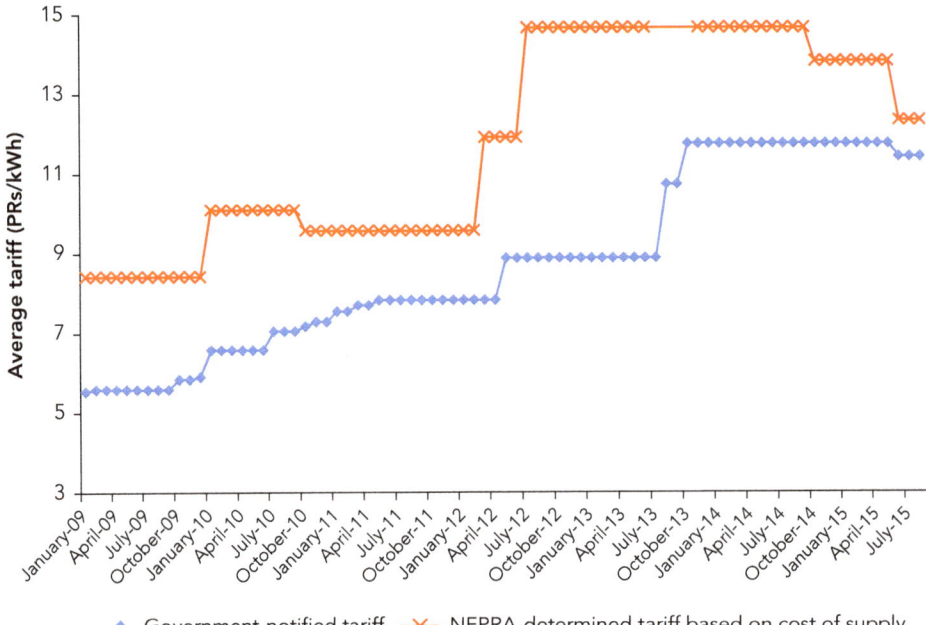

Source: World Bank calculation based on various tariff determinations and tariff notifications.
Note: The orange line indicates the average retail electricity tariff set by the National Electric Power Regulatory
Authority (NEPRA), based on the cost of delivering electricity to consumers. That cost includes the power purchase
cost, the targeted transmission and distribution cost, and a guaranteed return on the capital investments of
distribution companies. The blue line indicates the uniform tariff billed by the government. kWh = kilowatt-hour;
PRs = Pakistani rupees.

480 billion of the circular debt to breathe life into the paralyzed power sector. But the
underlying structural issues persist, and the circular debt has built up again, reaching
about PRs 414 billion by March 2017 (Ghumman 2017).

The tariff policy not only has a direct impact on electricity supply but also provides
weak incentives for distribution companies to reduce electricity losses and for consumers to conserve energy. NEPRA specifies targets for reducing transmission and distribution losses, but distribution companies failing to meet these targets face no repercussions
because there is an implicit guarantee of a government bailout—that is, the government
ultimately pays for their inefficiencies by clearing the circular debt. The tariff policy
fails to reward distribution companies for exceeding the targets because any resulting
savings are used to cross-subsidize poorly performing distribution companies. On the
demand side, electricity subsidies not only are regressive (Trimble, Yoshida and Saqib
2011; Walker and others 2014) but also encourage an entire set of inefficient behaviors
that exacerbate electricity shortages.

There are at least two effects of efficient pricing on welfare. First, if all distribution companies received a cost-recovery tariff, there would be no circular debt. Clearing up the circular debt would bring idled generation capacity—an estimated 5,000 MW in fiscal 2014—back into the system (World Bank 2015). The supply curve shifts downward because at any price there is more electricity output. Second, if the price were no longer fixed by regulators but could move until the market clears, there would be no unmet demand (implying a shift along the supply curve). Because of data limitations, the analysis only quantifies the second effect of inefficient pricing, that is the static effects of electricity underpricing. It therefore presents a lower-bound estimation of the cost of pricing distortion.

The exact welfare cost of electricity underpricing depends on the shapes and positions of the demand and supply curves. The more responsive supply and demand are to price, the larger is the deadweight loss from pricing below cost. Using the estimated supply and demand elasticities for electricity, the analysis projects that the new equilibrium quantity would be 9 percent higher than the current supply. Consumers lose from higher prices but producers would gain from increased supply and government would save on subsidies spending. The net welfare cost from electricity underpricing is estimated to be $363 million (about 0.13 percent of GDP) in fiscal 2015.

Downstream

Pakistan has made remarkable progress in connecting its towns and villages to the electric grid over the past few decades. The Global Tracking Framework report issued jointly by the World Bank and the International Energy Agency ranked Pakistan fourth in the world in terms of the number of people who gained access to electricity between 1990 and 2010 (after India, China, and Indonesia) (SEforAll 2014). Over this period, roughly 91 million people in Pakistan received electrical services for the first time.

The actual access rate to electricity in Pakistan is very much up for debate, however—and by any measure a large share of the population continues to live without electricity 24/7. Pakistan's latest official household survey reported that 97.5 percent of the population had access to electricity in 2016 (99.7 percent in urban areas and 95.6 percent in rural areas). Estimates based on census data and the number of connections reported by utilities suggest that access to grid electricity was only about 74 percent in 2016 (90 percent in urban areas and 63 percent in rural areas in 2016) (IEA 2017b). A 2014 survey sponsored by the International Finance Corporation (IFC) suggests that 35 percent of Pakistan's population lacked access to electricity in 2014 and that the access rate in parts of the country was alarmingly low (almost half of the population in Sindh lived off-grid, for example) (IFC 2015). Taken together, these estimates suggest that 5 million–54 million people in Pakistan may still lack access to grid electricity.

Pakistan also has the worst power outages in the region as measured by both duration and frequency. The 2013 World Bank Enterprise Survey found that 81 percent of firms in Pakistan reported being affected by outages, typically lasting 17 hours each. The 2014 IFC survey suggests that on average Pakistani households face 16 hours a day of load shedding during the summer and 12 hours a day during the winter (IFC 2015).

Power outages disrupt daily life for millions of households in Pakistan. They prevent people from using electric fans or pumping clean water in the sizzling heat of summer, when temperatures can reach 130°F. They make it difficult for students to study. They even prevent hospitals from performing operations. In frustration, people have taken to the streets to protest the failures in power delivery (Guardian 2012; Walsh and Masood 2013).

Power outages also affect business operations in Pakistan. They have forced hundreds of factories to downsize or shut down in the past few years, resulting in a drastic contraction of exports. Textile manufacturers, which account for more than half of Pakistan's export shipments, say that the frequent outages and long hours of load shedding make it difficult to meet order deadlines—and they report that buyers have shifted their business to countries such as Bangladesh and Vietnam (Mangi and Kay 2016). Small and medium-size factories suffer the most because they have fewer options for coping with power outages. Larger factories try to cope by investing in expensive captive power plants and diesel generators. Overall, more than three-quarters of firms identify electricity as a major constraint to their operation and growth, according to the 2013 World Bank Enterprise Survey.

Using detailed survey data, this section quantifies the welfare effects that electricity shortages have on households, firms, and the environment.

INSTITUTIONAL: WELFARE LOSS FOR HOUSEHOLDS

To quantify the costs incurred by households from lack of access to electricity, the analysis uses data from a two-period panel survey, the Pakistan Social and Living Standards Measurement Survey, for 2007/08 and 2010. The survey was carried out by the Pakistan Bureau of Statistics in all urban and rural areas of the four provinces and Islamabad, excluding restricted military areas. It collected information on a wide range of topics such as income, expenditure, education, health, water supply and sanitation, and electricity consumption. The first round, carried out during July 2007–June 2008, covered 15,512 households. The follow-up round, conducted during January–July 2010, covered about half the households surveyed in the first round. The sample size was smaller in the follow-up round because only households that could be interviewed in the same quarter of the year in which they were interviewed the first time were included. An analysis based on the historical data is used to identify the causal relationship between access to grid electricity and household welfare outcomes. The results are then used to estimate the cost of the lack of reliable access to electricity on households on the basis of the electrification rate in 2016.

Data from a second survey are used to gauge the impact of power outages on households. This survey, conducted in 2014 under the IFC's Lighting Pakistan Program, explored the market potential for off-grid solar lighting in Sindh and Balochistan in Pakistan. It included 6,000 households, both on- and off-grid, across the country's four provinces. In interviewing on-grid households, it asked about the daily average duration of outages. This cross-sectional survey lacks the multifaceted measurement of welfare supplied by the Pakistan Social and Living Standards Measurement Survey, but it does provide information on household income.

Because the Pakistan Social and Living Standards Measurement Survey reports an electrification rate of nearly 100 percent in urban areas, the analysis is restricted to the rural sample (4,300 households). According to the survey data, rural electrification is highest in the provinces of Khyber Pakhtunkhwa, followed by Punjab—both with a rate above 90 percent (Figure 5.21). Sindh and Balochistan have the lowest rates, at less than 75 percent in 2010. The average electrification rate rose by just 2 percentage points between 2008 and 2010, although the rate in Balochistan went up by about 17 percentage points and the rate in Sindh by 4 percentage points. The IFC survey reveals a much lower electrification rate in Sindh, where 50 percent of the population still lacked a connection to the grid in 2014.

An initial look at the data reveals striking differences in a wide range of welfare metrics between households with and without access to electricity (Figure 5.22). Households connected to the grid use substantially less kerosene than households that are not connected. They also have higher incomes and expenditures, a lower poverty rate, and better education outcomes. In addition, evidence suggests that electrification is associated with improved empowerment of women.

FIGURE 5.21 The electrification rate is still very low in some provinces of Pakistan

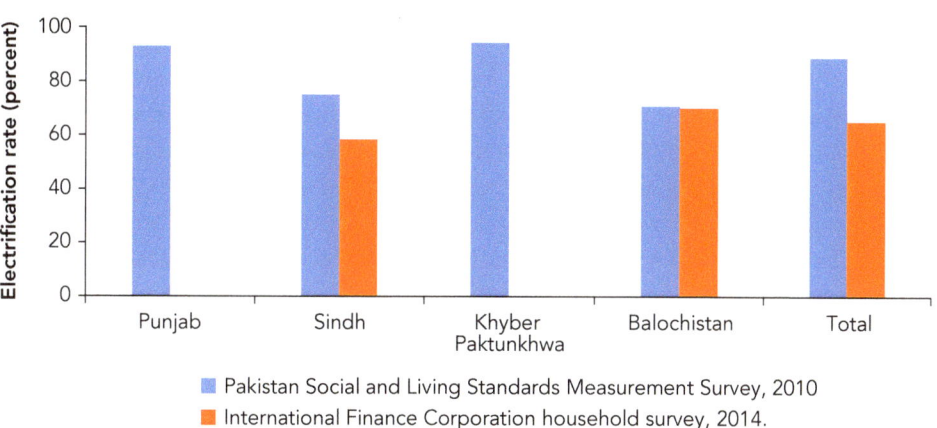

■ Pakistan Social and Living Standards Measurement Survey, 2010
■ International Finance Corporation household survey, 2014.

Source: Pakistan Bureau of Statistics (2010); IFC (2015).

FIGURE 5.22 Households with access to electricity have better welfare outcomes than households without access in Pakistan

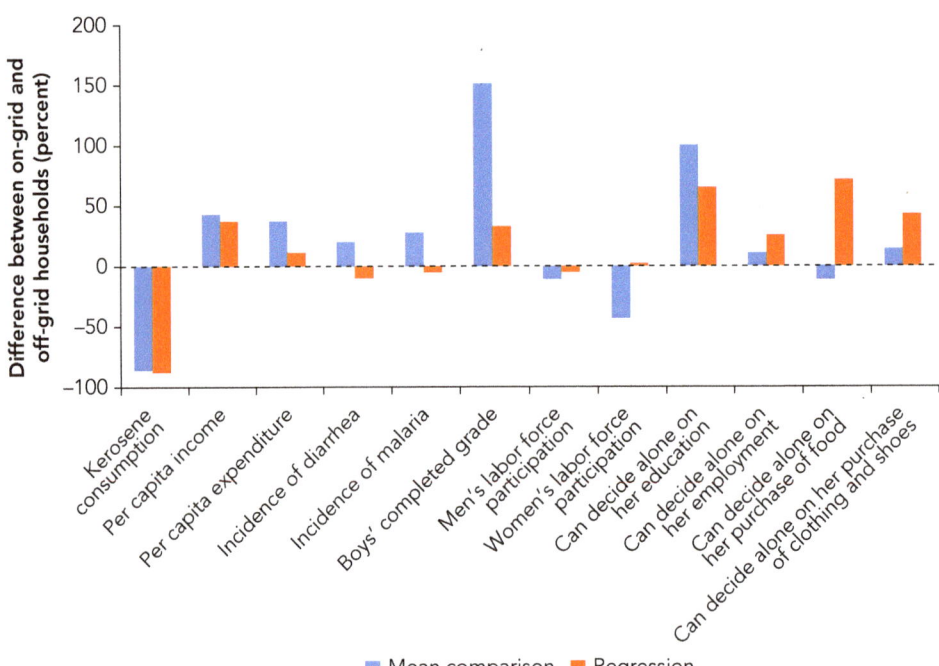

Source: Estimation based on Pakistan Bureau of Statistics (2008, 2010).
Note: Mean comparison refers to a simple average comparison without controlling for confounding factors. Regression refers to estimated difference based on econometric analysis.

To what extent does access to electricity cause these differences? An econometric analysis examines this question while taking into account the potential for simultaneous causality between electrification and welfare outcomes. This potential might arise, for example, because a government may direct electrification to areas that are easier to access or have greater growth potential—or because, when electricity becomes available in a village, households that are better able to afford it may connect to the grid earlier. Over time, households that are in more developed areas or were more well-off in the first place would achieve higher incomes even without electrification.

The analysis uses a two-stage propensity score–weighted fixed-effects model to control for unobserved village- and individual-specific effects that may simultaneously affect electrification status and the outcomes of interest. In the first stage, the analysis estimates a household's probability of being connected to the grid (propensity score), conditioned on a range of community and household characteristics observed in 2007/08. A weight variable (between 0 and 1) created from this propensity score is then applied to the full sample. The intuition in creating the weight variable is that households that more closely resemble those that eventually gained access to electricity receive higher weights in the sample (households with a grid connection in 2010 receive a weight of 1)—see Hirano, Imbens, and Ridder (2003). Samad and Zhang (2018) describe in detail the estimation strategy and results.

In the second stage, the analysis exploits the correlation between electrification and variations in welfare outcomes for the same household ("within" changes) between 2007/08 and 2010 while controlling for a range of household and community characteristics that could also affect this correlation. The characteristics include the age, gender, and education of the head of household; the number of adult men and women in the household; the amount of household agricultural land; and measures of household sanitation status such as access to running water and a flush toilet. They also include the village price of alternative fuels (firewood, kerosene) and of essential food items (staples, meat, fish, vegetables). The analysis controls for yearly shocks common to all households.

The results show that electrification indeed has a significant positive impact on households' welfare (see Figure 5.22). Gaining access to electricity led to a reduction of up to 88 percent in kerosene consumption and to an increase of 37 percent in per capita income and 11 percent in per capita expenditure. Connecting to electricity is also associated with better health outcomes, probably through access to electronic media, which improved knowledge of hygiene and healthy lifestyles, and through the reduction in kerosene consumption. In households that gained access to electricity, the incidence of diarrhea fell by 10 percent and the incidence of malaria by 5 percent among children under age 5.

Electrification of households also improved education outcomes for children and employment opportunities for women, although the effects are not widespread. Gaining access to electricity increased years of schooling completed by boys but had no effect on girls' education attainment. Electrification increased labor force participation for women but reduced it for men. The negative effect on men's labor force participation could be related to the effect of higher income on leisure.

Electrification increases women's decision-making power. In households gaining access to electricity, the probability of women making their own decisions rose by 25 percent for decisions on employment, 65 percent for decisions on education, and up to 71 percent for decisions on purchases. The explanation may lie in women's increased participation in the labor force (and thus greater economic empowerment), their increased access to electronic media (such as radios, televisions, and computers), or both.

Another important dimension of the welfare effects of electrification is the reliability of electricity supply. The Pakistan Social and Living Standards Measurement Survey reports only whether a household is connected to the grid, not how often it experiences outages. The IFC survey provides such data. Using data from the IFC survey, a basic regression analysis shows that each additional hour of daily outages is associated with an income loss of roughly 1.6 percent.

Reliable access to electricity is accompanied by a broad range of social and economic benefits. Not all of them can be quantified, although the potential gains in income growth alone are substantial. The monthly average income of rural households was

PRs 26,452 ($253) in fiscal 2014, according to the 2014 Household Integrated Economic Survey. With estimated average income gains of 37 percent a year, the per household monthly income gain would be about PRs 9,787 ($93). Assuming the marginal cost associated with electricity generation and transmission is about PRs 12.2 ($0.12) per kWh, annual average per capita electricity consumption is 471 kWh, and the average household includes 6.7 people, the net per capita gain from acquiring access to electricity is estimated at PRs 11,782 ($113) a year.

There is no consensus on the access rate of electricity in Pakistan. The official estimate based on household surveys suggests that about 5 million people remained off-grid. Data from the 2017 census and utility connections lead to an estimate that is almost 10 times as high. Using the more conservative figure of 5 million, connecting the entire off-grid population would raise income by $565 million a year. Using the higher figure, the annual income gain could reach $5.7 billion.

Improving the reliability of electricity supply would add to these gains. Anecdotal evidence suggests that power cuts have been reduced over the past few years, thanks to additional generation capacity and low global oil prices. Lacking official estimates of load-shedding hours, the analysis assumes that average load shedding was reduced to six hours a day in fiscal 2015. With an estimated income loss of 1.6 percent associated with every hour of daily outage, rural households would reap another $3.9 billion in annual income gains if electricity were provided 24/7.

The net income loss from lack of reliable access to electricity for households is therefore conservatively estimated at $4.5 billion (1.7 percent of GDP) a year. But this estimate likely grossly understates the actual loss because it does not capture the impact of unreliable electricity supply on health and education outcomes and because access rates could be much lower than the officially reported 97.5 percent.

INSTITUTIONAL: PRODUCTIVITY LOSS FOR FIRMS

Electricity outages impose sizable costs on firms. Without a reliable supply of electricity, firms must substitute away from energy-intensive capital or divert investments toward diesel generators—and credit constraints and market imperfections can compound these inefficiencies. In the absence of alternative sources of electricity, particularly for unanticipated outages, firms must send workers home, which reduces the productivity of labor.

To quantify the effects of electricity shortages on outcomes for firms, the analysis delves into data for 4,500 firms in 23 manufacturing sectors in Pakistan. These data come from two sources. The first is the Census of Manufacturing Industries, conducted by the Pakistan Bureau of Statistics (Bureau of Statistics Punjab 2010–11). It provides a thorough annual overview of firm-level activities, including detailed information on a variety of input costs such as labor, capital, and electricity, as well as revenue data. The analysis uses data from the 2010–11 census. This round mainly covered firms in Punjab, which account for most manufacturing activity in Pakistan.

The second source is NEPRA's *Performance Evaluation Report* on distribution companies (NEPRA 2011–15). This report publishes data on power shortages reported annually to NEPRA by each distribution company. The latest evaluation report includes shortage data for individual distribution companies from fiscal 2011 to fiscal 2015. Of the 11 distribution companies in Pakistan, 6 provide service in Punjab. The analysis identifies service areas by district for each distribution company and matches firm-level data with shortage data using district-level identifiers in the census data for 2010–11.

Shortages are measured by two reliability indexes and average load-shedding hours reported by distribution companies. The two reliability indexes are the System Average Interruption Frequency Index (SAIFI), which captures the average number of power supply interruptions that a customer experiences in a year, and the System Average Interruption Duration Index (SAIDI), which captures the average total duration of outages (in minutes) a customer experiences in a year. Both measures exclude outages caused by load shedding, but they do indicate the severity of unexpected outages.

There is large variation in the SAIFI and SAIDI values and in the average daily duration of load shedding across the six distribution companies that provide service in Punjab. An econometric analysis exploits the variation in the reliability of power supply to identify the relationship between shortages and firm productivity. Firm productivity is measured by revenue and value added per unit of inputs. The analysis also controls for a range of characteristics of firms such as labor costs, raw material costs, and total electricity costs, as well as sector-specific, time-invariant characteristics.

The analysis shows that a one-hour increase in the average daily duration of power outages leads to a reduction in a firm's value added and revenue of roughly 1.26 percent (Grainger and Zhang 2017a). The impact of unexpected outages measured by the SAIFI and SAIDI values is much larger. A one-hour increase in unexpected outages reduces firms' revenue by 9 percent on average.

When the average total duration of unexpected outages in a year is held constant, more frequent unexpected outages appear to be less detrimental, though the effect is insignificant. When the total duration of outages is held constant, more frequent outages in a year mean a shorter average duration of outages each time. This result suggests that on average firms may prefer more frequent but shorter unexpected outages to fewer but longer-lasting ones.

The effect of outages on productivity varies across sectors. The most energy- and technology-intensive manufacturers suffer the severest impacts. For manufacturers of various metal products, for example, the impact of outages on revenue and value added is one order of magnitude higher than the average impact across all sectors.

Although data used in the analysis do not cover the services sector, results from the 2013 World Bank Enterprise Survey suggest that electricity outages have an even greater impact on this sector. That survey provides information on 1,247 firms throughout Pakistan. Average losses in revenue from electrical outages were 29.6 percent for the manufacturing sector and 38.3 percent for the services sector in 2013.

To quantify the cost of power outages to firms, the analysis uses the more conservative estimates obtained in Grainger and Zhang (2017a). Total value added was about $33.9 billion for manufacturing and $133.7 billion for services in fiscal 2015. Assuming average load shedding is four hours a day for businesses, the loss of value added is estimated at $1.7 billion for manufacturing and $6.7 billion for services. The combined cost of power shortages on business is $8.4 billion (about 3.1 percent of GDP) a year.

One caveat about this analysis is that areas that provide a more business-friendly environment and experience faster economic growth may have a higher demand for electricity, which in turn could result in worse power shortages (Allcott, Collard-Wexler, and O'Connell 2016; Grainger and Zhang 2017b). The potential simultaneous causality between electricity shortages and economic growth implies that the results just described are likely to be lower-bound estimates of the impact of power shortages on firms.

SOCIAL: EMISSIONS FROM KEROSENE LIGHTING AND SELF-GENERATION

Another consequence of unreliable access to grid electricity is that it forces households and businesses to resort to other options, such as kerosene lamps and captive generators running on gas and diesel, to meet their energy needs. These options have harmful environmental effects. Kerosene lamps increase indoor air pollution, and captive generators, which are typically less fuel-efficient than conventional power plants, result in wasteful combustion of fuel and higher carbon dioxide emissions.

In Pakistan, about 4.7 million kerosene lamps are used by households and 600,000 lamps by businesses to serve basic lighting needs (Tedsen 2013). The use of kerosene lamps is not only a major source of indoor air pollution that has been linked to numerous health problems but also a source of ambient black carbon, the second-largest climate warmer in the atmosphere. Kurokawa and others (2013) estimate that black carbon emissions from kerosene lighting amount to about 0.06 gigagrams a year in Pakistan, equivalent to 53,300 tons of carbon dioxide emissions in warming effects. On the basis of a shadow price of carbon dioxide emissions of $40 per ton, the external costs of black carbon emissions from kerosene lamps in Pakistan are estimated at $2.13 million a year. Because lack of data makes it impossible to account for the health and safety cost of kerosene lamp use, this estimate understates the true social cost of kerosene lighting.

Another common way households in Pakistan cope with power outages is to use an uninterrupted power supply unit. Roughly 60 percent of households have installed some form of this device, according to an estimate by the Pakistan Environment Protection Agency (Shahid 2012). Although these units do not directly contribute to carbon dioxide emissions, they are not energy-efficient and increase indoor air pollution because their lead batteries emit poisonous fumes.

A large share of firms in Pakistan also rely on captive power generation to cope with power outages. The latest estimates indicate that more than 65 percent of businesses in Pakistan rely on self-generation, compared with an average of 45 percent in South Asia (World Bank 2013). In fiscal 2015, about 10 percent of the natural gas consumed in Pakistan went to the captive generation plants of industrial firms. Unlike gas-based power stations, which are typically located far away from population centers, captive units tend to be scattered across broad geographic areas. These captive plants have been estimated to have an average gas efficiency of only 18–28 percent, or much lower than the 35 percent average for gas-based power plants (Bhutta 2015). Captive generation is therefore likely to have greater environmental impacts, such as emissions of nitrogen oxide, than gas-based power stations.

Stand-alone diesel generators are also growing in importance in both the residential and commercial sector. In addition to being noisy, diesel generators release fumes containing more than 40 types of toxic air contaminants. Their proximity to population centers and prolonged use exacerbate the health and environmental risks they pose. Lack of data precludes an estimate of the cost of the use of diesel generators in Pakistan. The downstream social cost presented in this report therefore significantly underestimates the health and environmental costs imposed by an inefficient power system.

Summarizing the Cost

The total economic cost of distortions in the power sector in Pakistan is estimated to be $17.69 billion (about 6.53 percent of GDP) in fiscal 2015 (Table 5.1). The fiscal cost, consisting of consumer subsidies for electricity, is $2.15 billion (0.80 percent of GDP).

The impact of lack of reliable access to electricity on households and firms imposes the largest cost on the economy, estimated at $12.87 billion (4.75 percent of GDP) a year in fiscal 2015. It includes the potential income lost by the roughly 5 million people who still live off the grid and the millions of households and business that are affected by power outages (assuming average daily load shedding was six hours for rural households and four hours for business).

This figure is likely to greatly underestimate the actual cost of power shortages in Pakistan for at least two reasons. First, lack of reliable access to electricity has negative implications for a range of social and economic outcomes, such as educational achievement, health, and gender inequality. These losses are difficult to quantify and are not included in the calculation. Second, the number of people without access to grid electricity could be much higher than the conservative estimation of 5 million. According to estimation based on the 2017 census and the number of connections reported by utilities, 26 percent of the population—almost 51 million people in rural Pakistan—still has no access to grid electricity.

The second-largest sources of distortion include the inefficient allocation and delivery of gas and underinvestment in transmission. Each is estimated to cost about $1.1 billion (0.41 percent of GDP) a year. The power sector does not get highest priority in gas allocation. Diverting gas from fertilizer to power generation would increase electricity supply, reduce oil imports, and lower domestic fertilizer prices. Lowering UFG

TABLE 5.1 Cost of power sector distortions in Pakistan at a glance
percent of GDP

Type of cost	Upstream	Core Generation	Core Dispatch	Core Transmission	Core Distribution	Downstream	Total
Fiscal	0	0	0	0	0.80	0	0.80
Institutional	0.41	0.35	—	0.41	0.32	4.75	6.24
Regulatory	0.13	0	—	—	0.13	—	0.26
Social	0.03	0	—	—	—	0.001	0.03
Economic	0.57	0.35	—	0.41	0.45	4.75	6.53

Source: World Bank estimation.
Note: — = Not available. Estimation is for fiscal 2015.

during transmission and distribution would further reduce domestic gas shortages. Constraints on the transmission and distribution network due to poor infrastructure and underinvestment have prevented effective evacuation of power and are responsible for 29 percent of the electricity shortfall in fiscal 2015.

The third-largest distortion is inefficient electricity generation, which is estimated to cost Pakistan about $0.96 billion (0.35 percent of GDP) a year. Other large economic costs stem from inefficient electricity distribution, estimated at $860 million (0.32 percent of GDP) a year, and gas and electricity underpricing, each estimated at around $360 million (0.13 percent of GDP) a year. Reducing inefficiencies in generation and distribution would increase net electricity supply whereas removing energy subsidies would eliminate circular debt and send proper price signals for energy conservation.

The analysis applies generally conservative assumptions throughout. It also ignores some distortions—including out-of-merit dispatch of electricity, the social cost of electricity transmission and distribution, and the impact of electricity cross-subsidies on industry competitiveness—because of data limitations. The estimate therefore represents a lower bound of the actual cost of power sector distortions.

References

Ali, Mubarik, Faryal Ahmed, Hira Channa, and Stephen Davies. 2016. "Pakistan's Fertilizer Sector: Structure, Policies, Performance, and Impacts." IFPRI Discussion Paper 1516, International Food Policy Research Institute, Washington, DC.

Allcott, H., A. Collard-Wexler, and S. D. O'Connell. 2016. "How Do Electricity Shortages Affect Industry? Evidence from India." *American Economic Review* 106 (3): 587–624.

Alvarez, Ramón A., Stephen W. Pacala, James J. Winebrake, William L. Chameides, and Steven P. Hamburg. 2012. "Greater Focus Needed on Methane Leakage from Natural Gas Infrastructure." *Proceedings of National Academy of Sciences of the United States of America* 109 (17): 6435–40.

Averch, Harvey, and Leland L. Johnson. 1962. "Behavior of the Firm under Regulatory Constraint." *American Economic Review* 52 (5): 1052–69.

Awan, Umul. 2017. "Impact of Gas Reallocation on Power Generation in Pakistan." Background paper prepared for this report, World Bank, Washington, DC.

Bhutta, Zafar. 2015. "No Fuel Oil: New Power Plants to Run on Coal or LNG." *Express Tribune*, February 5. https://tribune.com.pk/story/833620/no-fuel-oil-new-power-plants-to-run-on-coal-or-lng.

Bureau of Statistics Punjab. 2010–11. "Census of Manufacturing Industries." Database. http://bos.gop.pk/cmi.

Chaudhry, A. 2016. "A Panel Data Analysis of Electricity Demand in the Pakistani Industrial Sector." *Energy Source, Part B: Economics, Planning, and Policy* 11 (1): 73–79.

Fabrizio, Kira, Nancy L. Rose, and Catherine D. Wolfram. 2007. "Do Markets Reduce Costs? Assessing the Impact of Regulatory Restructuring on US Electric Generation Efficiency." *American Economic Review* 97 (4): 1250–77.

Ghumman, Mushtaq. 2017. "Circular Debt Reaches Rs. 414 Billion." *Business Recorder*, March 5.

Grainger, C. A., and Fan Zhang. 2017a. "The Impact of Electricity Shortages on Firms: Evidence from Pakistan." Policy Research Working Paper 8130, World Bank, Washington, DC.

———. 2017b. "The Impact of Electricity Shortages on Micro- and Small-Enterprises: Evidence from India." Background paper prepared for this report, World Bank, Washington, DC.

Guardian. 2012. "Pakistan Power Cut Riots Spread as Politician's House Stormed." June 19.

Hausman, Catherine, and Ryan Kellogg. 2015. "Welfare and Distributional Implications of Shale Gas." Brookings Paper on Economic Activity, Brookings Institution, Washington, DC.

Health Effects Institute. 2017. *State of Global Air: A Special Report on Global Exposure to Air Pollution and Its Disease Burden*. Boston.

Hirano, Keisuke, Guido Imbens, and Geert Ridder. 2003. "Efficient Estimation of Average Treatment Effects Using the Estimated Propensity Score." *Econometrica* 71 (4): 1161–89.

IEA (International Energy Agency). 2017a. World Energy Balance and Statistics Database. https://www.iea.org/statistics/relateddatabases/worldenergystatisticsandbalances/.

———. 2017b. *Energy Access Outlook 2017 From Poverty to Prosperity*. World Energy Outlook Special Report. Paris.

IFC (International Finance Corporation). 2015. *Pakistan Off-Grid Lighting Consumer Perception Study*. Washington, DC.

Khan, Muhammad Arshad. 2015. "Modelling and Forecasting the Demand for Natural Gas in Pakistan." *Renewable and Sustainable Energy Reviews* 49: 1145–59.

KPMG. 2017. *Unaccounted for Gas Study*. Commissioned by Oil and Gas Regulatory Authority, Karachi, Pakistan.

Kurokawa, J., T. Ohara, T. Morikawa, S. Hanayama, G. Janssens-Maenhout, T. Fukui, K. Kawashima, and H. Akimoto. 2013. "Emissions of Air Pollutants and Greenhouse Gases over Asian Regions During 2000–2008: Regional Emission Inventory in Asia (REAS) Version 2." *Atmospheric Chemistry and Physics* 13 (21): 11019–58.

Mangi, Faseeh, and Chris Kay 2016. "Half a Million Jobs Lost as Textile Crisis Hits Pakistan's Economy." *Bloomberg*, September 21.

Ministry of Finance, Pakistan. 2003. *Pakistan Economic Survey 2002–03*. Islamabad.

———. 2013. *Pakistan Economic Survey 2012–13*. Islamabad.

———. 2015. *Pakistan Economic Survey 2014–15*. Islamabad.

———. 2017. *Pakistan Economic Survey 2016–17*. Islamabad.

Ministry of Petroleum and Natural Resources, Pakistan. 2005. *Natural Gas Allocation and Management Policy, 2005*. Division of Petroleum and Natural Resources, Islamabad.

Ministry of Petroleum and Natural Resources and Hydrocarbon Development Institute/Ministry of Energy. Various years. *Pakistan Energy Yearbook*. Islamabad.

Nasir, Muhammad, Muhammad Salman Tariq, and Arif Ankasha. 2008. "Residential Demand for Electricity in Pakistan." *Pakistan Development Review* 47 (4): 457–67.

NEPRA (National Electric Power Regulatory Authority). Various years. *State of Industry Report*. Islamabad.

———. 2011–2015. *Performance Evaluation Report*. Islamabad.

———. 2015. *Annual Report 2014–15*. Islamabad.

———. 2016. *Annual Report 2015–16*. Islamabad.

NTDC (National Transmission and Dispatch Company). 2014. *Electricity Demand Forecast Based on Regressions Analysis (Period 2008 to 2030)*. Government of Pakistan, Islamabad.

———. *2014. Electricity Demand Forecast Based on Multiple Regression*. Government of Pakistan, Islamabad.

———. 2016. *Power System Statistics*. Government of Pakistan, Islamabad.

OGRA (Oil and Gas Regulatory Authority). 2016. *Annual Report*. Islamabad.

———. 2017. *Annual Report*. Islamabad.

Pakistan Bureau of Statistics. 2008. Pakistan Social and Living Standards Measurement Survey 2007–08. Islamabad.

———. 2010. Pakistan Social and Living Standards Measurement Survey 2010. Islamabad.

———. 2014. Household Income and Expenditure Survey 2013–14. Islamabad. http://www.pbs .gov.pk/sites/default/files//pslm/publications/hies2013_14/tables/TABLE_11_2014.pdf.

Pakistan State Oil. 2018. HSFO and HSD Price Archives (database). Islamabad. http://psopk .com/en/product-and-services/product-prices/hsfo/hsfo-archive.

Parry, Ian, Dirk Hein, Eliza Lis, and Shanjun LI. 2014. *Getting Energy Prices Right: From Principle to Practice*. Washington, DC: International Monetary Fund.

Samad, Hussain, and Fan Zhang. 2018. "Electrification and Household Welfare: Evidence from Pakistan." Policy Research Working Paper 8582, World Bank, Washington, DC.

Sanchez-Triana, Ernesto, Santiago Enriquez, Javaid Afzal, Akiko Nakagawa, and Asif Shuja Khan. 2014. *Cleaning Pakistan's Air: Policy Options to Address the Cost of Outdoor Air Pollution*. World Bank, Washington, DC.

Schwab, Klaus. 2018. *The Global Competitiveness Report 2017–2018*. Geneva: World Economic Forum.

SEforAll (Sustainable Energy for All). 2013. *Global Tracking Framework*. Vienna.

Shahid, Jamal. 2012. "UPS and Generators Heavy on Pocket—and Health Too." *DAWN*, June 4. https://www.dawn.com/news/723767.

Tedsen, E. 2013. *Black Carbon Emissions from Kerosene Lamps: Potential for New CCAC Initiative.* Berlin: Ecological Institute.

Trimble, Chris, Nobuo Yoshida, and Mohammad Saqib. 2011. *Rethinking Electricity Tariffs and Subsidies in Pakistan.* Washington, DC: World Bank.

USAID (U.S. Agency for International Development). 2011. *Evaluation of Economic Value of Natural Gas in Various Sectors.* Washington, DC.

———. 2013. *The Causes and Impacts of Power Sector Circular Debt in Pakistan.* Study commissioned by the Planning Commission of Pakistan. Washington, DC.

Walker, Thomas, Sebnem Sahin, Mohammad Saqib, and Kristy Mayer. 2014. *Reforming Electricity Subsidies in Pakistan: Measures to Protect the Poor.* World Bank Policy Paper Series on Pakistan, PK 24/12, Washington, DC.

Walsh, Declan, and Salman Masood. 2013. "Pakistan Faces Struggle to Keep Its Lights On." *New York Times*, May 27.

World Bank. Various years. World Development Indicators (database). Washington, DC.

———. 2003. *Pakistan Oil and Gas Sector Review.* Washington, DC.

———. 2013. Enterprise Survey. Washington, DC. http://www.enterprisesurveys.org.

———. 2015. "Pakistan: Second Power Sector Reform Development Policy Credit Program Project Document." Washington, DC.

Zhang, Fan. 2007. "Does Electricity Restructuring Work? Evidence from the U.S. Nuclear Energy Industry." *Journal of Industrial Economics* 55 (3): 397–418.

CHAPTER 6

Conclusion

U sing a common analytical framework, the report quantifies the monetary value of policy-induced distortions at every stage of the power supply in Bangladesh, India, and Pakistan. The results of this analysis reveal that the economic cost of these distortions is much greater than previously thought. Comprehensive energy sector reform aimed at enhancing incentives for efficiency and quality and correcting pricing distortions in both the fuel and electricity markets could play a substantial part in boosting living standards and economic growth in South Asia.

Costs Are Much Higher than Previously Thought

Focusing on economic costs, the analysis finds that power sector distortions are far greater than previously estimated on the basis of their fiscal costs alone. The total economic costs were about $11.2 billion (about 5.0 percent of the gross domestic product, GDP) in Bangladesh in fiscal 2016; about $86.1 billion (4.1 percent of GDP) in fiscal 2016 in India; and $17.7 billion (6.5 percent of GDP) in fiscal 2015 in Pakistan. In Bangladesh, the underpricing of gas is the largest source of the economic cost of distortions, responsible for an annual loss of $4.5 billion (2.0 percent of GDP). In India, the environmental effects of excessive coal use are the largest source of cost, estimated at $35.4 billion (1.7 percent of GDP) a year. In Pakistan, the impact of the lack of reliable access to electricity on households and firms is the largest source of the economic cost, leading to annual losses of roughly $12.9 billion (4.8 percent of GDP).

These results suggest the importance of making power sector reform a top priority. Indeed, few other reforms could yield such large economic gains as quickly. Moreover, by expanding access to electricity and improving the quality of supply, power sector reform would directly benefit poor households. The highest payoffs are likely to come from institutional reforms, a focus on expanding reliable access, and the appropriate pricing of the negative health and environmental impact of emissions from fossil fuel-based power generation.

Distortions May Reinforce or Offset One Another

Distortions in the power sector often amplify or offset one another. Institutional distortions tend to reinforce regulatory distortions and vice versa. For example, inefficient production and allocation in the input and output markets for electricity have led to higher electricity costs and a greater need for subsidies to keep end prices low. The use of subsidies makes it difficult to differentiate losses from utility mismanagement from losses from unpaid subsidies, weakening the accountability of utilities for their performance.

Institutional and regulatory distortions at different stages of the power supply amplify one another's effects. For example, fuel allocation based on the ownership of utilities rather than on their efficiency undermines competition in electricity generation. In addition, both institutional and regulatory distortions in the upstream and core sectors contribute to downstream power shortages and poor quality of electricity supply.

The relationship between social and other types of distortions is mixed. On the one hand, regulatory distortions generally increase the cost of social distortions. The underpricing of energy, for example, leads to wasteful consumption and excessive emissions or the depletion of groundwater. On the other hand, inefficient energy production may partially offset social distortions by limiting energy supply.

This interaction across distortions means that their combined effects could be smaller or greater than the sum of the parts from a partial equilibrium analysis. Data limitations preclude a general equilibrium analysis that considers all potential interactions between distortions, however. Results presented in the report nonetheless provide a first order approximation of the total cost of distortions in the power sector.

The potential interaction between distortions also has important policy implications: policy reforms that correct one distortion while leaving others intact may have the unintended consequence of exacerbating losses elsewhere. Particularly important is avoiding a narrow focus on liberalizing the price of electricity, because in the absence of institutional reforms, the market equilibrium is highly inefficient. Without improvements in efficiency, pricing reform is also more politically challenging. Failure to improve service could undermine the sustainability of the pricing reform and even lead to its reversal. Similarly, without complementary pricing reform, efforts to reduce institutional distortions can have perverse effects. Increasing access to and the supply of electricity, for example, can lead to higher emissions or more groundwater pumping.

All this suggests that, when reforms are comprehensive, they can be more sustainable and yield greater benefits. Also critical when tackling distortions that are interrelated is addressing them directly at their source (Bhagwati 1971). As suggested by the celebrated theorem of Jan Tinbergen (1956), to achieve the first-best optimum one must have as many policies as there are distortions.

Regional Distortions Hinder Greater Cross-Border Electricity Trade

This analysis focuses on internal distortions and their impacts on the domestic power market. However, regional distortions that prevent more cross-border electricity trade impose large opportunity costs as well. South Asia could play a large role in the cross-border electricity trade. Differences in seasonal patterns of energy supply and demand make a case for exchanging electricity across countries to address electricity shortages cost-effectively. Furthermore, the enormous potential for hydropower in Nepal and Bhutan, estimated at more than 40 gigawatts (GW) in Nepal alone, can be developed only if there is access to larger markets.

Many studies have discussed the potential benefits of increased electricity connectivity within South Asia. Those benefits include economies of scale and scope in investments, greater renewable energy development and a reduced dependence on fossil fuel imports, and enhanced competition in electricity supply, among others (ESMAP 2010; Singh and others 2013; Srivastava and Misra 2007; UNESCAP 2016). Timilsina and others (2015) quantify the potential gains from full regional trade in electricity in South Asia—that is, an unrestricted flow of electricity between any two parts of the region as determined by the markets. They estimate the benefits for the period 2015–40, measured against a baseline scenario that incorporates all existing policies and assumes that each country makes its own capacity investments independently. They find that during the study period full regional trade in electricity would result in net annual fuel savings (net of investment cost) of about $9 billion and a reduction of carbon dioxide emissions of about 8 percent. With access to a greater regional market, a significant number of hydropower plants will be built, and roughly 50 GW of coal power plants will be replaced as a result.

Regional trade in power provides large potential for enhancing the electricity supply and reducing emissions, but much of this potential remains unexploited. The current cross-border cooperation in electricity has been restricted to bilateral trade arrangements, mainly with India, which imports about 1.5 GW of hydropower from Bhutan. India also exports about 600 MW of electricity to Bangladesh and around 190 MW to Nepal (Government of India 2017; Press Information Bureau 2017).

Various political, institutional, and regulatory distortions have contributed to the current low level of cooperation in the electricity sector. Singh and others (2015) have thoroughly analyzed these distortions. They find that lack of trust and historical animosity between countries have obstructed efforts to enhance regional cooperation, including in electricity. Although there has been some enthusiasm for a regional power market, those trying to achieve this goal have been frustrated because of internal political conflicts and lack of political will.

The absence of regulatory coordination across countries to ensure harmonized rules for transmission access, system operation, congestion management, energy accounting, data transfer, and so on is also a problem. The electricity trade is also affected by levies, such as export, import and transit taxes. Although the South Asian Free Trade Agreement envisioned a common regional market, it did not give special treatment to the electricity trade.

Internal regulatory distortions create impediments to the cross-border electricity trade as well. Pricing electricity below cost-recovery levels reduces incentives to create the generation and transmission capacity needed to expand cross-border trade. Policies favoring incumbent utilities also create barriers to entry for new players in a regional market. Developing a well-functioning domestic power market would therefore facilitate greater cross-border cooperation and trade in electricity.

Implications for Power Sector Reform

In the past, power sector reform has often focused on the traditional power segments—generation, transmission, and distribution—with the primary goal of attracting private investment to the sector. Installing new power plants, poles, and lines is important for meeting the rapidly increasing demand for electricity, but reforming the upstream fuel supply and improving the productivity of existing infrastructure are also needed to achieve reliable, affordable, and sustainable service delivery. Institutional reforms offer a cost-effective way to eliminate a large part of existing power shortages because of the potential for achieving large efficiency gains by strengthening incentives for improving managerial performance. By contrast, without fundamental changes in incentive structures, corporatizing power utilities does not guarantee meaningful improvements in their operating performance.

Pricing reform also needs to be an integral part of a reform package. Pricing energy below its private and social cost discourages production, generates excess demand and emissions, and creates perverse incentives for distribution utilities to underserve loss-making customers. Combined with electoral incentives that reward short-term, more visible investment over long-term maintenance efforts, below-cost pricing inevitably creates tension between the quantity and quality of electricity connections. But removing electricity subsidies means a sharp increase in energy prices. To buffer the impact of price hikes, efficiency needs to be increased rapidly on both the supply and demand side. Meanwhile, targeted social assistance should be provided to the people most affected.

LOOK BEYOND THE CORE SECTOR

To achieve a reliable and sustainable electricity supply, governments need to look beyond the traditional power segments to address distortions in the upstream fuel sector. These distortions have contributed to growing shortfalls in the domestic fuel supply, forcing power plants to shut down or operate below capacity. Fuel shortfalls also must be met by expensive imports of liquid fuel, resulting in high-cost generation and worse emissions. In Bangladesh, for example, fuel constraints have reduced the gas-fired generating capacity by more than 10 percent. In Pakistan, the reliance on imported oil has partially contributed to a colossal circular debt, debilitating the entire supply chain of power. Pricing fossil fuel below its social cost also causes excessive emissions and health damage. As this analysis shows, upstream distortions have caused some of the greatest economic waste in the countries.

To secure an adequate fuel supply, it is imperative to reform the coal and gas sectors to improve the efficiency of production and delivery. Doing so requires measures to open fuel markets to private entrants, introduce effective competition, and limit government's political interference in day-to-day operations. In India, the government has already taken actions to commercialize coal mining. Various other measures are needed to introduce competition into an otherwise monopolistic coal market. Policy changes could spur investment in automated technologies to improve the efficiency and safety of coal mining. Strong implementation to optimize linkages between mines and plants and actions to ease cross-subsidization in freight tariff could also improve coal delivery.

In Pakistan, efforts to reduce high-level gas losses in distribution are needed urgently. In the absence of full privatization, those efforts should focus on limiting the government's political interference in operation and investment, implementing rigorous monitoring, and enforcing performance standards. Removing obstacles to wholesale competition would encourage new entrants and add competitive pressure for incumbents to raise operational efficiency.

Pricing reform could also play an important role. Evaluated by their opportunity costs, fuel subsidies are much larger than electricity subsidies in South Asia. Charging prices that reflect the full economic cost of fuel has multiple advantages. It provides greater incentives for upstream production, sends price signals to curtail demand, and facilitates efficient allocation of resources across multiple sectors because consumers who can extract more economic value from fuel are willing to pay more for it. Pricing fuel to reflect its social cost generates net benefits by reducing environmental and health damage. In India, more than two-thirds of these benefits would arise from the local air pollution avoided.

When it is not feasible to balance supply and demand through pricing, fuel should be allocated on the basis of efficiency rather than favoritism. Allocating fuel to public enterprises or sectors with political connections without concern for efficiency undermines the productivity of fuel use. Not allowing generators to compete for fuel on an equal footing can also deter private investment in the power market.

Overreliance on a single fuel for power generation raises reliability concerns because anything that restricts the availability of that fuel could have serious implications for the cost and reliability of the power sector. Achieving a diverse fuel portfolio, with different types and sources of fuel, requires a holistic approach. Countries will need to engage more in regional energy cooperation and scale up the development of previously untapped renewable resources.

THINK BEYOND INVESTMENT

Investment is urgently needed in some segments of the power sector. However, investment by itself is unlikely to solve the problem of power shortages in South Asia because of the big role played by inefficiency. Ranging from low productivity in generation to high losses in distribution, inefficiency exacerbates power shortages. Addressing inefficiency, often through policy changes rather than investment, is the most cost-effective way to reduce shortages because it would maximize the use of existing infrastructure.

A key factor is the quality of managerial practices, which have a direct effect on utility performance. Analysis based on plant-level production data reveals that large and persistent variations in productivity remain even after controlling for differences in technology, vintage, consumer mix, and other physical and exogenous characteristics of utilities within a country. These variations in productivity are attributable to innate differences in firm operation. For power plants, a nontrivial factor is ownership type: private power stations are up to 30 percent more efficient than public ones in turning fuel into electricity. For distribution utilities, limited variation in ownership makes causal analysis challenging. But, within a group of largely poorly managed distribution utilities, the best performers in India and Pakistan meet international standards, suggesting that levers within management's control can lead to big gains in distribution performance.

Institutions define the incentive structure for firms and their managers, and there is growing empirical evidence that competition and private participation can play a key part in encouraging managerial effort. Competition can be promoted in a range of ways in different segments of the power sector, including by ensuring nondiscriminatory access to fuel for public and private producers alike, dispatching generation in merit order from lowest to highest variable cost, and removing discriminatory charges levied on consumers buying electricity on the open market. Apart from outright privatization, other ways to tap private sector initiative include franchise arrangements in electricity distribution and contracts to outsource system operations and maintenance. For example, to reduce the tendency of distribution companies seeking bailout packages from the government, steps can be taken towards privatization of management control. In this effort, it may be vital to introduce an expert panel recommended by regulatory energy commissions composed of private experts to manage the power plant and not taking government's bailouts as a failure buffer.

In the absence of market-based pricing for energy, the use of incentive-based regulation can boost operational efficiency. In the South Asian power sector, tariff setting for generation and for gas and electricity transmission and distribution largely follows rate-of-return regulation. Production costs are passed on to consumers, and utilities receive premiums for capital-related costs. Traditional rate-of-return regulation provides incentives for capital investment but not for efficient operation or high-quality service.

Energy regulators around the world have used different kinds of incentive mechanisms to replace traditional rate-of-return regulation in order to improve the performance of utilities. The general idea is through tariff setting to apply penalties for falling below performance standards and offer rewards for exceeding them. The performance standards can be predetermined, based on industry norms, or periodically adjusted, based on the relative performance of similar utilities within a country.

Some incentive schemes, such as multiyear tariff regulation and the linking of distribution tariffs to performance targets, have already been adopted in South Asia. This kind of regulation requires rigorous benchmarking, monitoring, and performance evaluation. Analysis presented in this report provides an example of the type of statistical benchmarking required. For state-owned enterprises, this regulatory approach also requires governments to link budget to performance—that is, to remove soft budget constraints.

To get the biggest bang out of investment dollars, investment needs to be targeted. Targeting is especially important in a decentralized environment with growing participation by private investors. In the transmission sector, for example, implementing location-based pricing that links transmission charges with the opportunity cost of congestion could help channel investment to the parts of the network most affected by congestion.

REFORM BEYOND CORPORATIZATION

Corporatization of state-owned enterprises has been a key government strategy for power sector reform in South Asia. This kind of reform is aimed at improving the performance of public utilities with no change in ownership, increasing their managerial autonomy, and holding them to a higher standard of accountability. In Bangladesh, 40 percent of public generation utilities have been corporatized. In India, 86 percent of distribution utilities (by sales value) have been corporatized.

Despite the popularity of corporatization, evidence for its success (in the absence of privatization) is limited. At first glance, corporatized utilities in Bangladesh and India appear to be more efficient than utilities under direct public control. When differences in technical characteristics are taken into account and performance is evaluated over time, however, the advantages of corporatized entities disappear. In Bangladesh, the fuel productivity and total factor productivity of corporatized generation utilities are statistically indistinguishable from those of power plants under direct government control (see Figure 3.16). In India, corporatized distribution utilities have lower technical and commercial losses and higher collection efficiency than utilities managed by the state power department. But the corporatized utilities were more efficient before corporatization. Analysis focusing on within-firm changes reveals that utilities corporatized during 2007–13 actually experienced higher losses after corporatization. The increase was largest in the first year but persisted over time.

This evidence suggests that, without parallel reforms, corporatization alone may not guarantee improved operating efficiency. To succeed, corporatization must be supported by effective incentive and monitoring mechanisms. The Power Grid Company of Bangladesh and the Dhaka Power Distribution Company cut transmission and distribution losses substantially after corporatization, in part because their shares are traded on the local stock market. The stock market plays a unique role in monitoring and rewarding managerial performance (Holmstrom and Tirole 1993).

Ensuring an arm's-length relationship between corporatized entities and the government is also important for effective corporatization. Because the government remains the controlling owner, corporatized utilities may remain susceptible to political interference in employment and pricing policies. The government's pursuit of objectives that conflict with efficiency goals can distort the constraints and incentives faced by managers.

If corporatized utilities are expected to behave like private entities, governments must impose hard budget constraints. Unlike privately owned utilities, which must recoup fixed investment costs through tariffs, publicly owned companies may recover costs through government subsidies, such as those implied by soft budget constraints. Repeated government bailouts would undermine incentives to reduce losses for any form of enterprise, corporatized or not.

PRIORITIZE QUALITY, NOT JUST ACCESS

Achieving universal access to electricity should remain high on the governments' agendas. Almost 255 million people in South Asia still lack electricity. Providing them with access would bring them a broad range of social and economic benefits, as the analysis described in this report documents. But merely ensuring connectivity is not enough. The gains from electrification depend critically on whether the "connected" households receive an adequate level of service (Samad and Zhang 2016, 2017, 2018). Electrical wires alone provide no benefit if there is no power. Without power to provide light, children cannot study in the evenings and businesses cannot remain open in the evening. Poor-quality electricity service also discourages sustained changes in study and work patterns—and even the adoption of electricity in the first place (Banerjee and others 2015).

The electrification rate has risen in South Asia, but the increase has not necessarily been associated with an increase in the availability of electricity. Indeed, extension of the grid may have caused a decline in the overall quality of electricity service. In India, many villages officially classified as electrified remained in the dark for years after the completion of electrification projects (Gaba and Min 2016). Analysis based on high-frequency nighttime lights data for India in 2013 suggests that power cuts were positively correlated with new electrification projects. Districts with more recently electrified villages suffered from worse subsequent power outages in 2013. And within a district, outages were worse in previously electrified villages when more neighboring villages were electrified in the previous period (see Box 4.3).

Both regulatory and political imperfections can lead to a trade-off between the quantity and quality of electricity connections. Where electricity prices are too low to recover costs, adding new connections inevitably increases the strain on the grid because the system is forced to absorb more loss-making customers. And when cash-strapped distribution utilities have limited resources or incentives to cope with greater demand for electricity, the quality of the electricity supply suffers, affecting both new and existing customers.

Electoral incentives can cause distortions, too. Politicians may favor short-term, more visible investments in grid extension over long-term, hidden investments in the software, operations, and maintenance important for reliable service. In a budget-constrained environment, this drive toward quantity often comes at the expense of quality, resulting in subpar infrastructure for the vast majority (Scott and Seth 2013).

Policy makers can act in various ways to move toward more inclusive and accountable service delivery. To ensure that distribution utilities have the financial resources they need for investing in and maintaining the grid, policy makers can remove electricity subsidies and increase revenue collection. Tariffs that recover costs also eliminate perverse incentives to underserve loss-making customers.

Engaging citizens in monitoring service delivery can also be a powerful tool for improving its quality. By allowing citizens to prioritize and monitor public spending, politicians can build trust in and understanding of government reform decisions.

Also critical is improving the collection and sharing of data on power outages. Understanding where and whose power gets cut improves accountability. Utilities tend to underreport load shedding and are often reluctant to share outage data at the district or village level. Data from high-frequency satellite imagery of nighttime lights allow monitoring of power supply disruptions in close to real time (Box 4.3). Such data offer an alternative for regulators and power system planners seeking power outage data at a highly disaggregated level.

ACCOMPANY REFORMS WITH COMPENSATION

Removing regulatory and social distortions requires adopting energy prices that reflect the true cost of energy consumption. Doing so involves both removing subsidies to bring energy prices closer to market rates and internalizing the social costs of energy consumption into prices. Although price reform delivers large economic and environmental benefits in the long term, hikes in energy prices can cause immediate economic distress, especially for the poor and vulnerable. Raising prices gradually while providing targeted social assistance, for example, through direct benefit transfer, can mitigate the impact.

Subsidies need to be phased out gradually, following a preannounced schedule. Managing public expectations about future energy price changes is important because uncertainty could hurt both business spending and individual welfare (Pindyck 1991). Moreover, households and firms will respond to higher energy prices by either improving the efficiency of energy use or substituting away from energy. A gradual approach would allow them to smooth out adjustment costs over time. By contrast, a sudden jump in price could provoke strong social and political opposition, undermining the sustainability of reform.

Even with a slow approach, increasing energy prices could still raise concerns about affordability. Higher energy prices affect consumer purchasing power both directly, through higher energy costs, and indirectly, through higher prices for goods and services for which energy is an input. The burden could fall most heavily on poorer households. They typically spend a larger share of income on energy and energy-intensive products such as food (Grainger and Kolstad 2010).

Offsetting price increases requires rapid improvements in efficiency on both the supply and demand side. Improving efficiency in generation would lower electricity costs and relieve upside pressure on prices. Improving efficiency in end use could reduce total energy consumption (assuming limited rebound effects). To help low-income households increase energy efficiency, many countries have provided assistance aimed at improving the affordability of energy conservation measures (such as switching to energy-efficient appliances) or funded education programs offering practical information on simple ways to save energy (Deichmann and Zhang 2013). This type of efficiency program provides long-term support for energy affordability without recurrent expenses.

Efficiency programs may, however, take time to have an effect. To protect the poor in the immediate term, it is important to assess potential poverty and social impacts before reform and provide targeted social assistance to the people most affected—either by scaling up existing programs or implementing new ones. An effective social assistance program would balance the need for coverage with the need for targeting. The choice typically includes lifeline tariffs and direct cash transfers, whether or not earmarked for energy consumption (Vagliasindi 2013). India, for example, introduced a Direct Benefit Transfer Scheme—the Pratyaksha Hastaantarit Laabh (PAHAL) program to facilitate the removal of price-distorting subsidies for liquefied petroleum gas (LPG) in a socially sensitive way. Under the program, LPG cylinders are sold at the market price to households and subsidy on LPG cylinders is credited directly to consumers' bank accounts after they have purchased gas cylinders. The program helped to avoid market distortions and diversion of subsidized LPG to non-subsidized sectors that would result from a dual pricing mechanism. The program also avoided hard-hitting impact on poor households through direct cash transfer.

In general, there is no "one size fits all" formula. Instead, the design and implementation of complementary social assistance should build on lessons from general good practices in tariff reform while taking into account the national or even local context.

A price on emissions would also prompt countries to move toward renewables and away from fossil fuel–powered electricity. Although new jobs and opportunities are created during the process, workers in and communities reliant on the fossil fuel industry could experience massive social and economic disruptions, including unemployment, poverty, and fragmentation. These impacts should not be overlooked. Complementary policies are needed to ensure a just transition.

International experience offers numerous examples of a shift to a sustainable economy with limited adverse impacts on workers. A well-structured plan typically includes retraining programs and strategies for pursuing greater diversification in the local economy. These programs help miners and other workers in polluting industries gain new skills and provide job placement, job matching, and options for work exposure during the transition. Creating new long-term local economic opportunities helps regions and communities thrive and should be an important part of the jobs program. Many countries have set up special transition funds to provide affected workers with direct financial compensation and cover the costs of education and job training.

References

Banerjee, Sudeshna Ghosh, Douglas Barnes, Bipul Singh, Kristy Mayer, and Hussain Samad. 2015. *Power for All: Electricity Access Challenge in India*. Washington, DC: World Bank.

Bhagwati, Jagdish. 1971. "The Generalized Theory of Distortions and Welfare." In *Trade, Balance of Payments and Growth*, edited by J. N. Bhagwati, R. W. Jones, R. A. Mundell, and J. Vanek. Amsterdam: North-Holland.

Deichmann, Uwe, and Fan Zhang. 2013. *The Economic Benefits of Climate Action*. Washington, DC: World Bank.

ESMAP (Energy Sector Management Assistance Program). 2010. "Regional Power Sector Integration: Lessons from Global Case Studies and a Literature Review." Briefing Note 004/10, World Bank, Washington, DC.

Gaba, Kwawu, and Brian Min. 2016. *Tracking Electrification in India Using Night Time Lights*. Washington, DC: World Bank.

Government of India. 2017. "Cross Border Electricity Trade" (web tool). http://indiaenergy.gov.in/iess/supply_eleimport.php.

Grainger, C. A., and C. D. Kolstad. 2010. "Distribution and Climate Change Policies." In *Climate Change Policies: Global Challenges and Future Prospects*, edited by E. Cerda and X. Labandiera. Cheltenham, U.K.: Edward Elgar Publishing.

Holmstrom, Bengt, and Jean Tirole. 1993. "Market Liquidity and Performance Monitoring." *Journal of Political Economy* 101: 678–709.

Min, Brian, Zachary O'Keeffe, and Fan Zhang. 2017. "Whose Power Gets Cut? Using High-Frequency Satellite Images to Measure Power Supply Irregularity." Policy Research Working Paper 8131, World Bank, Washington, DC.

Pindyck. Robert. 1991. "Irreversibility, Uncertainty, and Investment." *Journal of Economic Literature*. 29 (3): 1110–48.

Press Information Bureau. 2017. "India Becomes Net Exporter of Electricity for the First Time." Ministry of Power, Government of India, New Delhi. http://pib.nic.in/newsite/PrintRelease.aspx?relid=160105.

Scott, Andrew, and Prachi Seth. 2013. *The Political Economy of Electricity Distribution in Developing Countries: A Review of the Literature*. London: Overseas Development Institute.

Samad, Hussain, and Fan Zhang. 2016. "Benefits of Electrification and the Role of Reliability: Evidence from India." Policy Research Working Paper 7889, World Bank, Washington, DC.

———. 2017. "Heterogeneous Effects of Rural Electrification: Evidence from Bangladesh." Policy Research Working Paper 8102, World Bank, Washington, DC.

———. 2018. "Electrification and Household Welfare: Evidence from Pakistan." Policy Research Working Paper 8582, World Bank, Washington DC.

Singh, Anoop, Tooraj Jamasb, Rabindra Nepal, and Michael A. Toman. 2015. "Cross-Border Electricity Cooperation in South Asia." Policy Research Working Paper 7328, World Bank, Washington, DC.

Singh, Anoop, Jyoti Parikh, K. K. Agrawal, Dipti Khare, Rajiv R. Panda, and Pallavi Mohla. 2013. "Prospects for Regional Cooperation on Cross-Border Electricity Trade in South Asia." Integrated Research and Action for Development, New Delhi.

Srivastava, Leena, and Neha Misra. 2007. "Promoting Regional Energy Co-Operation in South Asia." *Energy Policy* 35 (6): 3360–68.

Timilsina, Govinda R., Michael A. Toman, Jorge G. Karacsonyi, and Luca de Tena Diego. 2015. "How Much Could South Asia Benefit from Regional Electricity Cooperation and Trade?" Policy Research Working Paper 7341, World Bank, Washington, DC.

Tinbergen, Jan. 1956. *Economic Policy: Principles and Design.* Amsterdam: North-Holland.

UNESCAP (United Nations Economic and Social Commission for Asia and the Pacific). 2016. *Promoting Regional Energy Connectivity in Asia and the Pacific.* Bangkok.

Vagliasindi, Maria. 2013. *Implementing Energy Subsidy Reforms: Evidence from Developing Countries.* Directions in Development. Washington, DC: World Bank.

Methodology for Estimating Demand and Supply Elasticities

This appendix describes the methodology used to estimate the short-run and long-run demand and supply elasticities for fuel and electricity. To simplify the discussion, in this appendix it is assumed that price is the sole determinant of demand and supply. (In the report, the methodology is generalized to include multiple determinants of demand and supply, including income and the prices of substituting fuels.)

Let y_t^d and y_t^s denote the demand and supply of a good (coal, gas, or electricity) at time t. The relation between y_t and p_t (both in log form) can be expressed as the following:

$$y_t^d = \beta_d p_t^d + \varepsilon_t^d$$
$$y_t^s = \beta_s p_t^s + \varepsilon_t^s$$

where p_t^d and p_t^s are consumer and producer prices, ε_t^d and ε_t^s are shocks to demand and supply, and β_d and β_s are price elasticities. Price and quantity usually trend over time. Careful analysis for the presence of unit root and cointegration is needed to avoid spurious regression in which the regression may be picking up a relationship between the trends in two variables rather than an underlying relationship between the variables.

The autoregressive distributed lag model (ARDL) tests for cointegration. The lags of dependent variables as well as other predictors are used to control for serial correlation.

This appendix was contributed by Ashish Rajbhandari.

Pesaran, Shin, and Smith (2001) provide bounds test of cointegration in this framework that do not require pretesting for unit roots in individual series.

The ARDL model for the demand equation is specified

$$y_t^d = \gamma_0 + \sum_{i=1}^{p} \gamma_i y_{t-i}^d + \sum_{j=0}^{q} \beta_j p_{t-j}^d + \varepsilon_t^d \tag{A.1}$$

where γ_0 is the intercept, γ_i is the coefficient on lagged demand, and β_j is the coefficient on price. If the two series are cointegrated, there exists a long-run relation between demand and prices. Both series share a common trend, and any deviation between the series exists only temporarily.

The ARDL model can be transformed into an error correction model useful for obtaining short-run and long-run elasticities,

$$\Delta y_t^d = \gamma_0 + \left(\sum_{i=1}^{p} \gamma_i - 1 \right) y_{t-1}^d + \left(\sum_{j=0}^{q} \beta_j \right) p_{t-1}^d - \gamma_2 \Delta y_{t-1}^d - \dots - \gamma_p \Delta y_{t-p+1}^d$$

$$+ \beta_0 \Delta p_t^d - \beta_2 \Delta p_{t-1}^d - \dots - \beta_q \Delta p_{t-q+1}^d + \varepsilon_t^d$$

$$\Delta y_t^d = \gamma_0 + \left(\sum_{i=1}^{p} \gamma_i - 1 \right) \left[y_{t-1}^d - \left(\frac{\sum_{j=0}^{q} \beta_j}{1 - \sum_{i=1}^{p} \gamma_i} \right) p_{t-1}^d \right] - \gamma_2 \Delta y_{t-1}^d - \dots - \gamma_p \Delta y_{t-p+1}^d \tag{A.2}$$

$$+ \beta_0 \Delta p_t^d - \beta_2 \Delta p_{t-1}^d - \dots - \beta_q \Delta p_{t-q+1}^d + \varepsilon_t^d$$

where the short-run elasticity is β_0 and the long-run elasticity is $\dfrac{\sum_{j=0}^{q} \beta_j}{1 - \sum_{i=1}^{p} \gamma_i}$.

To test for the existence of cointegration, Pesaran, Shin, and Smith (2001) suggest estimating the parameters of equation (A.2) with the unrestricted error correction model

$$\Delta y_t^d = \alpha_0 + \theta_y y_{t-1}^d + \theta_p p_{t-1}^d + k_{y,2} \Delta y_{t-1}^d + \dots + k_{y,p} \Delta y_{t-p+1}^d + k_{p,0} \Delta p_t^d + \dots + k_{p,q} \Delta p_{t-q+1}^d + \varepsilon_t.$$

The null hypothesis for the test of no cointegration is H_0: $\theta_y = \theta_p = 0$. The restriction is tested by constructing an F-test statistic. The critical values are obtained from the table of bounds in Pesaran, Shin, and Smith (2001).

For the price and quantity series analyzed in this report, cointegration is established between supply/demand and prices. The analysis then estimates coefficients of equation (A.2) (and price elasticities) using the generalized method of moments. Price and quantity are often simultaneously determined. To control for the potential endogeneity of price, the analysis instruments price with either its own lags or the lags of other instruments. The length of the lag is selected on the basis of Akaike Information Criteria. The analysis computes standard errors using the Heteroskedastic and Autocorrelation Consistent estimator with Bartlett kernel.

Reference

Pesaran, M. H., Y. Shin, and R. Smith. 2001. "Bounds Testing Approaches to the Analysis of Level Relationships." *Journal of Applied Econometrics* 16: 289–326.

Use of the Stochastic Production Frontier Approach to Measure the Technical Efficiency of Utilities

T his appendix describes the general framework of the stochastic production frontier approach used to analyze the technical (in)efficiency and total factor productivity of power plants and distribution utilities described in the report. The approach was first proposed by Aigner, Lovell, and Schmidt (1977) and Meeusen and van den Broeck (1977) and further investigated by Kumbhakar and Lovell (2000).

A recent development of stochastic frontier analysis is the modeling of the effects of environmental variables (z)—ownership types in this analysis. These exogenous or policy variables may affect technical inefficiency.[1] Kumbhakar and Sun (2013) propose a generalized framework in which inefficiency and the noise term are a function of z. This analysis applies their approach. The environmental variables appear in both the pretruncation mean and the variance of inefficiency as well as in the variance of the noise term. The parameters are estimated using the maximum likelihood estimator in one step.

To better understand the approach, consider the following stochastic production frontier

$$Y = f(X; \beta)\exp(v\text{-}u)$$
$$= f(X; \beta)\cdot\exp(v)\cdot TE \qquad (B.1)$$

This appendix was contributed by Kai Sun.

where Y is observed output. For power plants, Y is total electricity output in log; for distribution utilities, Y is total electricity sold in log. X is an input vector, including labor, capital, and operations and maintenance in logs and possibly some other exgenous factors, such as time trend and consumer mix. β is a vector of technology parameters, v is noise, $u \geq 0$ is the production or technical inefficiency, and

$$TE = \exp(-u) = \frac{Y}{f(X;\beta)\exp(v)} \tag{B.2}$$

is the technical efficiency (TE) score, where $f(X;\beta)\exp(v)$ is interpreted as the maximum feasible output. The TE score measures the ratio of actual output to maximum feasible output. Because $u \geq 0$, by definition $0 \leq TE \leq 1$.

If $f(X;\beta)$ takes the log-linear Cobb-Douglas form, then

$$\log Y = \beta_0 + \sum_{j=1}^{k} \beta_j \log X_j + v - u. \tag{B.3}$$

Defining $y = \log Y$, $x = [1 \ \log X]'$, $\beta = [\beta_0 \beta_1 ... \beta_k]'$ and adding subscripts i (for either plant or distribution utility) and t (for time period) to (B.3) yields

$$y_{it} = \beta'x_{it} + v_{it} - u_{it}. \tag{B.4}$$

As before, v_{it} is the production noise and u_{it} the production or technical inefficiencies. Following Kumbhakar and Sun (2013), u_{it} is specified as

$$u_{it} \sim i_{id} N^{+}(\mu_{it}, \sigma_{uit}^{2}), \tag{B.5}$$

with a pretruncation mean of

$$\mu_{it} = c_0 + \delta'z_{it}, \tag{B.6}$$

and a pretruncation standard deviation of

$$\sigma_{uit} = \exp(c_1 + \gamma'z_{it}). \tag{B.7}$$

Then, v_{it} is specified as

$$v_{it} \sim i_{id} N(0, \sigma_{vit}^{2}), \tag{B.8}$$

where

$$\sigma_{vit} = \exp(c_2 + \rho'z_{it}). \tag{B.9}$$

and u_{it} and v_{it} are independently distributed of each other and of the regressors. The environmental variables (ownership types), z_{it}, can affect technical inefficiency, u_{it}, through its pretruncation mean, μ_{it} as well as pretruncation variance, σ_{uit}^2. Noise, v_{it}, is considered heteroskedastic conditional on z_{it}.[2] The maximum likelihood estimator is then used to estimate the parameters in (B.4) (β, c_0-c_2, δ, γ, and ρ).[3] The log-likelihood function to be maximized for the ith plant (utility) in year t is

$$L_{it} = -\frac{1}{2}\ln\sigma_{it}^2 + \ln\left[\phi\left(\frac{y_{it} - \beta'x_{it} + \mu_{it}}{\sigma_{it}}\right)\right] - \ln\left[\Phi\left(\frac{\mu_{it}}{\sigma_{uit}}\right)\right] + \ln\left[\Phi\left(\frac{\tilde{\mu}_{it}}{\sigma_{*it}}\right)\right], \quad (B.10)$$

where $\sigma_{it}^2 = \sigma_{uit}^2 + \sigma_{vit}^2$ and $\sigma_{*it} = \sigma_{uit}\sigma_{vit}/\sigma_{it}$. Let $\varepsilon_{it} = v_{it} - u_{it}$ and $\tilde{\mu}_{it} = (\mu_{it}\sigma_{vit}^2 - \varepsilon_{it}\sigma_{uit}^2)/\sigma_{it}^2$ (Kumbhakar and Lovell 2000). After the parameters in (B.10) are estimated, the Jondrow, Lovell, Materov, and Schmidt (JLMS) (1982) estimator is used to compute the point estimates for technical inefficiency,

$$E(u_{it} \mid \varepsilon_{it}) = \tilde{\mu}_{it} + \sigma_{*it}\frac{\phi(\tilde{\mu}_{it}/\sigma_{*it})}{\Phi(\tilde{\mu}_{it}/\sigma_{*it})}. \quad (B.11)$$

The formula devised by Battese and Coelli (1988) is used to calculate the TE scores,

$$TE = E[\exp(-u_{it}) \mid \varepsilon_{it}] = \frac{\Phi(\tilde{\mu}_{it}/\sigma_{*it} - \sigma_{*it})}{\Phi(\tilde{\mu}_{it}/\sigma_{*it})} \cdot \exp(-\tilde{\mu}_{it} + \frac{1}{2}\sigma_{*it}^2), \quad (B.12)$$

where ϕ and Φ denote the standard normal density and distribution functions, respectively.

Notes

1. Simar, Lovell, and van den Eeckaut (1994) propose that inefficiency can be expressed as a standard half-normal random variable multiplied by a positive function of z. Other studies suggest a similar method of expressing the pretruncation mean or variance of inefficiency as a function of z (Reifschneider and Stevenson 1991; Caudill, Ford, and Gropper 1995; Wang 2002; Wang and Schmidt 2002).

2. Kumbhakar and Sun (2013) show that the z_{it} can affect the estimator of technical inefficiency of Jondrow and others (1982) through the conditional heteroskedasticity of v_{it} even if u_{it} is not a function of z_{it}, because the estimator is based on u_{it} as well as v_{it}.

3. z_{it} is a vector in general, hence δ, γ, and ρ are also vectors.

References

Aigner, D. J., C. A. K. Lovell, and P. Schmidt. 1977. "Formulation and Estimation of Stochastic Frontier Production Functions." *Journal of Econometrics* 6 (1): 21–37.

Battese, G. E., and T. J. Coelli. 1988. "Prediction of Firm-Level Technical Efficiencies with a Generalized Frontier Production Function and Panel Data." *Journal of Econometrics* 38 (3): 387–99.

Caudill, S. B., J. M. Ford, and D. M. Gropper. 1995. "Frontier Estimation and Firm-Specific Inefficiency Measures in the Presence of Heteroskedasticity." *Journal of Business and Economic Statistics* 13 (1): 105–11.

Jondrow, J., C. A. K. Lovell, I. S. Materov, and P. Schmidt. 1982. "On the Estimation of Technical Inefficiency in the Stochastic Frontier Production Function Model." *Journal of Econometrics* 19: 233–38.

Kumbhakar, S. C., and C. A. K. Lovell. 2000. *Stochastic Frontier Analysis*. Cambridge: Cambridge University Press.

Kumbhakar, S., and K. Sun. 2013. "Derivation of Marginal Effects of Determinants of Technical Inefficiency." *Economic Letters* 120 (2): 249–53.

Meeusen, W., and J. van den Broeck. 1977. "Efficiency Estimation from Cobb-Douglas Production Functions with Composed Error." *International Economic Review* 18 (2): 435–44.

Reifschneider, D., and R. Stevenson. 1991. "Systematic Departures from the Frontier: A Framework for the Analysis of Firm Inefficiency." *International Economic Review* 32 (3): 715–23.

Simar, L., C.A.K. Lovell, and P. van den Eeckaut. 1994. "Stochastic Frontiers Incorporating Exogenous Influences on Efficiency." Discussion Paper 9403, Institut de Statistique, Université Catholique de Louvain, Louvain-la-Neuve, Belgium.

Wang, H.-J. 2002. "Heteroscedasticity and Non-Monotonic Efficiency Effects of a Stochastic Frontier Model." *Journal of Productivity Analysis* 18: 241–53.

Wang, H.-J., and P. Schmidt. 2002. "One-Step and Two-Step Estimation of the Effects of Exogenous Variables on Technical Efficiency Levels." *Journal of Productivity Analysis* 18: 129–44.